MW01598338

The Mobile Internet Handbook

2015 U.S. RVers Edition

by

Chris Dunphy & Cherie Ve Ard

with guest author Jack Mayer

Third Edition: February 2015

Second Edition: August 2014
First Edition: June 2013

ISBN-13: 978-1508484493

Disclaimer: This is complicated stuff, and there are no easy, one-size-fits-all solutions. In this book, we share what we've learned over the years in our own journeys using extensive amounts of mobile internet along the way, as well as what we've learned from wide-ranging research and conversations with other mobile internet experts.

We have no formal affiliation or financial stake in any of the products or services we mention, except our own mobile apps. With anything you purchase, you are entering into transactions directly with the manufacturers and providers of those services.

This book is a sharing of our research and experience as full-time RVers ourselves. We can take no responsibility for the choices you make as a result of reading this book. We will do our best to share the pros and cons of each option as we understand them today, but ultimately you must continue your research and decide for yourself.

When issues arise, please seek support and resolution from the provider, vendor, or manufacturer you purchased from – not us!

Also by Technomadia:

No Excuses: Go Nomadic!

www.technomadia.com / excuses

A practical guide to the logistics of full-timing – income sources, mail, banking, healthcare, family, pets, safety, relationships, preparing, and much more!

The subject of this book is constantly evolving.

We're constantly staying on top of this topic, and a newer edition of this book may even be available by the time you read this. In between book updates, we post RV mobile internet relevant news as it comes out, as well as provide more in-depth guides, product reviews and more at:

www.RVMobileInternet.com

We welcome you to join our **free monthly e-mail newsletter** - we'll send you a summary of what's changed in the mobile internet landscape recently:

eepurl.com/0KJG1

Join our free public Facebook group for discussions with other RVers, ourselves included, interested in this topic:

www.facebook.com/groups/rvinternet

Premium Membership & Private Advising

We offer a premium membership service, called *Mobile Internet Aficionados*, for those who want to keep up to date on this stuff with exclusive in-depth content, news analysis, webinars, and private Q&A forums. It's our classroom.

We also offer advising sessions for those who'd like extra help figuring out their setup.

See the last chapter, The Ongoing Conversation, for more information and a money saving coupon for having bought this book already.

Dedicated to

Tim VeArd
1944 – 2013

Technology pioneer, national hero, and location-independent technology entrepreneur who inspired us in ways beyond imagination.

We miss you, Dad.

And a HUGE Thank You to all of our supporters who crowdfunded the rewrite and massive expansion of this book!

Table of Contents

Prologue to the 3rd Edition

In the spring of 2013, we set out to write a comprehensive blog post about all the options for keeping online while being mobile – bringing together in one place the years of content we had written on the topic.

By the time we were partway through, we knew this couldn't be covered in a single post – there was way more than a book's worth that could be written on the topic!

And thus, on the spur of the moment, the first edition of *The Mobile Internet Handbook* was born. We had never published a book before – and we put together the first edition in under 3 weeks on a shoestring budget, eager to get back to our real jobs.

We did not expect the book to take on a life of its own – gathering so many reviews and endorsements as a must read for RVers who need to keep online. Eventually we came to realize just how needed this sort of resource was.

Technology marches on, and by mid-2014 it was time to update the book.

But we didn't just want to quickly update the book to make it current, we also wanted to expand the content and provide more resources to help more people live vibrant and connected mobile lives. And this would take more than a couple weeks to accomplish!

To gauge just how much interest there was in a second and expanded edition, we decided to try a book pre-sale via the crowdfunding site Indiegogo.

We hoped to at least cover the costs of professional graphics, illustrations, and editing so that our new book would look more like a book and less like a long blog post. We set

1

several stretch goals for frequently asked-for topics that we could also add to the content. We expected at most to be on the hook for writing only a couple new chapters.

We were blown away by the support we received, and in the end every single one of our stretch goals was fully funded!

The second edition thus contained *seven* entirely new chapters, as well as vastly expanded content in all the rest. The original book was only 87 pages, and the second edition was well more than double that size.

We hired professionals to handle the editing and graphics, and the creation of our new RVMobileInternet.com website. We not only hired professionals, but fellow *mobile* professionals who are also dependent on the very topic of this book.

We're pleased as punch to work with Jeanette and Dennis of MotorhomeOfficeOfDesignAndTechnology.com for the style and design (from the cover art to the fun illustrations you'll see throughout the book, as well as the color and typography), Jason and Kristin of BoondockMarketing.com for the bones of the new website, and Ann of Beardsley Editing for the editing of this book.

We asked all of them for fair market bids for their services, but we know each one of them went above and beyond to help us make this an awesome resource for our mobile communities…and we can't recommend or thank them enough.

We're also thrilled that Jack Mayer, a fellow mobile tech guru we've always respected, approached us to be involved. He contributed a chapter to this book, will soon publish an advanced addendum book on Customer Premises Equipment (CPE) devices (get on our e-mail list at eepurl.com/0KJG1 to be notified when it's ready), and he has become a regular contributor of reviews and exclusive content to our RVMobileInteret.com website as well.

Now this brings us to this new third edition for 2015.

Things in the mobile internet world seem to never slow down. The new edition hadn't even been out for 2 months, and several major changes in the industry occurred. 2014 ended with a new mobile internet landscape - and we rose to the challenge of keeping up with it all through our new premium membership service and resource center.

Competition between the carriers has increased, new offers and plans have emerged, and new technologies have come to market. Further disruption came when our former top cellular recommendation for heavy data users,

Millenicom, vanished from the scene in October. Meanwhile, Verizon made a major policy changes concerning obtaining and keeping grandfathered unlimited plans. And more recently, some carriers added in long wished for features, like rollover data.

Thanks to the magic of modern publishing, we have been able to dive back into the book to create a refreshed 2015 edition - incorporating all the latest information, options, and recommendations - and polishing out some of the information we previously wrote.

If you have already read the 2nd edition - much will be familiar. This is not an epic rewrite and expansion like last year. But whether you skimmed or devoured the 2014 edition, you are certain to learn a thing or two this year.

We set out to create the definitive resource for RVers wanting to stay connected while on the road. We hope that you will agree that we have.

An Extra Thank You to Our Book Supporters

Listed below are the names of supporters who gave extra in advance to help make this book possible. We are eternally humbled and grateful for the funding we received so that we could take the time off from our other projects to focus on the research and writing of this book – and launching a new web resource for all: www.RVMobileInternet.com

A Technomadia Fan!
Andy & Wendy Young
Ann Beardsley & Elliott Walsh
Chuck & Ruth Treadway – OnTheRoad
Forrest & Mary Clark, Two For(e)Traveling
Fritz Neumann
George & Rhonda Kaupas
Harry & Tiffany Kidder
JMJ & LEJ
Joe & Wendy of AmericanGypsyGibberish.com
John Sasinowski
Joseph and Rachel Jones
Ken & Jackie Osinski – RViewoftheUSA.blogspot.com
KentWilliamsPhotography.com
Kiki Kikinator Potatohead
M. Proulx
Michelle Cegon
Mobile First
Molly McD and Family
Patricia A. Bowen, Ph.D. a.k.a. Villagetart
Randy Britt
Rev. Michael Hollinger, wanna-be Nomad
Rich & Deavon
Sean Mahoney of SeanPatrickMahoney.com
Steven and Linda of thechouters.blogspot.com
Susan, A Full-Timer and Technomadia Fan
The Copeland Odyssey
Tom & Stacie of RVTexasYall.com
Veronica and Denny – RVOutlawz.com
Vic Neshyba, Austin TX
William Bailey – TheJourneyVisvi.com

Introduction

More than likely the internet plays at least some role in your life – and for many of us, it is central.

For anyone thinking of hitting the road, figuring out how to best keep *online* while exploring the world *offline* becomes critically important.

Whether you use the internet to keep in touch with loved ones, navigate to your next destination, look up accommodations, manage your finances, learn online, pursue entertainment, or depend on connectivity for your income source – building a connectivity arsenal that suits you is an essential chore most techno-connected nomads face.

Even those who don't consider themselves techno-savvy at all still face needing at least some connectivity these days.

Who This Book Is and Isn't For

This book is focused on internet connectivity options for mobile folks based in the USA.

And more specifically, some of the resources are geared specifically towards RVers and those living on the road – whether full-time or seasonal.

If you are setting out to explore the vast expanse of the USA for a prolonged period of time and want to remain connected, you've found a

5

great resource. We also include a primer on staying connected while crossing borders into Canada, Mexico, and beyond.

We are attempting to write this book in a clear manner that will be accessible to everyone – ranging from folks who don't yet understand the difference between WiFi and cellular data, to those who are super geeks like us who understand the differences intuitively and who get turned on by advanced options.

However, there is no denying that at the root of it all, *The Mobile Internet Handbook* is pretty technical stuff; if you're not tech-savvy already, hopefully this book will help you bridge the gap to be able to understand at least the gist of the issues you'll face staying online.

Keep in mind, you're going to be out there on your own needing to manage whatever connectivity toolkit you assemble, often miles away from any geek help. And if you are having trouble getting online, you may end up unable to ask for assistance in forums and groups. So before you get lost in the wilds (figuratively or literally!), make sure you understand what you have and how to use it!

Read these chapters over and over again if you need to, and do further research. You may need to enlist the assistance of a geeky friend or family member to help you assemble and install your arsenal.

Don't be afraid to ask for help – before you really need it!

Who Is Technomadia?

We're Chris and Cherie of Technomadia.com.

We've been traveling full time since 2006, mostly via various RV setups. Currently we call a 1961 geeked-out and solar-powered vintage bus conversion our home, but we have also full-timed in small travel trailers in the past.

As we're both currently in our early 40s, we're too young to be traditionally retired and not fortunate enough to be retired early…yet. And we have some serious wanderlust.

Thankfully mobile technology enables us to take our high-tech careers in software development, strategy advising, and technology consulting on the

6

road, allowing us to work remotely for our clients while also creating our own products.

Our business is Two Steps Beyond LLC (www.twostepsbeyond.com).

We consider ourselves to be "technomads" and have been able to create a lifestyle that combines our careers and our desire for mobility.

Needless to say, we absolutely depend on mobile internet to keep connected to our clients, manage our software projects, and keep in touch with loved ones.

And well, we're just geeks who like to spend a lot of time online. Heck, we both pretty much grew up with being online as part of our teen social life in the 1980s, and we even met online (on a Prius forum) shortly after Chris first hit the road in 2006.

Before going nomadic, Chris worked in Silicon Valley, focused on mobile technology.

One of his jobs had him serving as the Director of Competitive Analysis (aka "Chief Spy") for Palm and PalmSource – the companies behind the pioneering Palm Pilot and Treo. His job was to be intimately familiar with every mobile device and technology in existence, and he was tasked with traveling the world to dig up information to chart the future of the entire mobile industry.

It was not uncommon for him to be carrying dozens of mobile devices with him at a time – always raising eyebrows passing through airport security scans.

Times haven't changed much actually – we still have dozens of mobile devices on board!

Cherie has run a software development business from home since the mid-1990s with a long history of working remotely for her clients. Her career involved technical writing and teaching high-tech topics to nontechnical people. When she met Chris during his first year on the road, she was accustomed to carrying smartphones able to tether her laptop to the internet. And it was essential for her to keep connected if she was to join him full-time.

Our first year on the road together was spent in a tiny teardrop travel trailer (16') equipped with just the essentials – solar electricity and mobile internet. Luxuries like air conditioning, refrigeration, and even plumbing had to wait a year until our second travel trailer (17') was custom built for us.

After so many years on the road, we have tried a wide variety of mobile internet solutions.

We know this stuff intimately, our lives are living laboratories for mobile technology, and it is our honor to share our wealth of experience with you.

Follow our life on the road at:

Blog: www.technomadia.com

Facebook: www.facebook.com/technomadia

YouTube: www.youtube.com/user/TalesFromTechnomadia

Be in touch at: contact@technomadia.com

About Our Guest Author, Jack Mayer

Jack Mayer has been a full-time RVer since 2000. He is a freelance writer specializing in RV-related technical topics, author of a popular RV-related website (www.jackdanmayer.com), and speaker at RV rallies. His professional background is in computer system software design, networking, and operating systems.

For the past 15 years, he has designed and implemented a variety of wired and wireless networks for RV parks, small businesses, and individuals. In addition to his work in the communications field, Jack has specialized in the design and implementation of RV electrical systems for off-grid living.

More About This Book

Just some disclosures:

- Except where specifically disclosed, we are not financially affiliated with any of the products or companies we talk about in this book, and we're not writing this book to sell you anything. We're just sharing our lessons learned, our tech backgrounds, and our years of experience living on the road while keeping connected. We strive to be as unbiased as we can.

- Yes, some of our tech equipment has been provided at no charge by the manufacturers/vendors – but these usually come with the expectations of us being reviewers or beta test sites in which we provide feedback to the companies to improve their products. We get a say in how these products evolve to better meet the needs of folks like us who are truly mobile. And being able to get free or loaner gear helps us review far more products without having to charge an arm and a leg for this book to cover equipment expenses. But some stuff we do buy on our own as well.

  If you know of any cool mobile technology that you think we should consider featuring in a future book update or on RVMobileInternet.com, let us know. If you are a manufacturer, let us know if you would like to send us some gear to test and review.

  Beware: You can count on our honest (sometimes brutally so) feedback.

- Much of our current tech is Apple products, though in the past we were both Windows users. We are not hopeless Apple fanatics – we have been won over by the quality of the hardware, software, and customer support – and now we have iPhones, iPads, a Mac Mini, and MacBook Pro laptop.

  However, despite our Apple-centric personal arsenal, this book it not platform specific – all are welcome here! We'll try to keep our Apple enthusiasm balanced.

- In addition to our consulting projects, we also write mobile travel apps for the iPhone, iPad, and Android.

'Coverage?' (www.twostepsbeyond.com/apps/coverage) is the app most related to this book, and you'll see screen captures in the book used to compare the various carriers. This powerfully simple app uniquely displays overlayable coverage maps from the major carriers to help us technomads know where we can best keep connected.

If you do buy that app, you *are* tossing a couple extra bucks into our account – thank you!

- Please don't assume any prices or plans mentioned in this book are necessarily current – this stuff changes often. This book isn't intended to be a price comparison guide, and we only use prices as examples as they are relevant today. It's almost a guarantee that as soon as we publish this book, some of those details will be immediately outdated.

- Although we do provide some technical and installation advice, please always consult with the product's vendor or manufacturer for direct support.

Armed with the information in this book, you will be much better equipped to understand and evaluate the current offerings on the market – and to decide what plans and technologies personally fit you best.

Our goal here is not to give you a shopping list for the singular perfect mobile internet setup – it is to arm you with the information you need to write your own.

Laying the Groundwork

First off – if at any point you come across any terms in this book that are at all unfamiliar to you, stop and check the glossary at the end!

We have written a comprehensive glossary that defines even the most technical terms in ways that anyone can understand. So before you get frustrated wondering why you might need a POE to power your CPE to get remote 802.11g when you'd really rather have more dB on your LTE – check the glossary, and soon it will all make sense.

Mobile Internet Options

11

Basic Differences Between the Common Options

There are multiple ways to access the internet while on the go these days, and each of them has attributes that make it more attractive than the others.

Public WiFi hotspots are often free or low cost, but they can vary vastly in quality and are frequently too overloaded to be a reliable source. Unless you're willing to take your laptop closer to the physical hotspot, you may need additional gear to get a usable signal from the comfort of your RV.

Cellular data is quite prevalent and has gotten amazingly fast – and is now even available in sometimes surprisingly remote locations. However, it's typically priced by how much you use, which can add up fast. You might need extra equipment or boosting gear to optimize utilizing cellular in remote locations. You'll need to select your carrier(s) and equipment wisely to best match your planned travel destinations and routes.

Sometimes signal just simply is not available though, especially as you leave populated areas.

Satellite can be picked up anywhere with access to the southern sky, even remote locations and across borders into Mexico and Canada – but satellite internet comes with a host of drawbacks, including speed (or lack of it), price, latency, and complexity.

More than likely, most RVers will create a personal arsenal that combines multiple options.

Here's a quick grid that shows the benefits of the primary ways to getting online:

	Unlimited or Capped	Mobile Friendly	Cost	Speed	Reliability
Cable/DSL	(Usually) Unlimited	Not mobile. Fixed location.	Reasonable	Fast!	Always on
Cellular	Capped	**Fully mobile – wherever there's signal!**	Pricey	Slow to fast (Faster all the time!)	Varies by location
Public WiFi	Variable	You hunt signal at each location	**Free to cheap**	Highly variable	Highly variable
Satellite	Capped	**Fully mobile – wherever there's southern sky**	Pricey to beyond pricey!	Slow and high latency	Can be persnickety

12

Cellular vs. WiFi

By far the two most common ways that RVers will be able to get online is via either a cellular data connection or via campground or another public WiFi network.

But – one of the most basic questions we occasionally get asked – just what exactly is the difference?

We'll be going into much more detail about both cellular and WiFi later in this book, but for those of you starting at square one, here is the basic breakdown:

WiFi – This is a short-range local wireless network technology that is built into all modern laptops, smartphones, and tablets. Via WiFi, you can connect to a nearby hotspot that is at most just a few hundred feet away (without special equipment). In general, WiFi connections tend to be free, but there are some paid ones. A hotspot may be one you create and host yourself, or it may be one provided by a campground, cafe, library, or hotel. The hotspot itself is providing access via whatever its own upstream internet source is – cable, DSL, satellite, or even cellular.

Cellular Data – This is a longer range data connection established over the same basic wireless network that carries cellphone voice calls. All smartphones, some tablets, and a very few laptops have cellular data capabilities built in. When using cellular data, you are talking to the internet via a cell tower that is typically up to 5 miles away, but in a remote area the tower may be 15 to 20+ miles away, and in a busy urban area there may be cell towers located every few blocks, or even inside buildings. Cellular data is rarely free.

Sitting squarely between cellular and WiFi are mobile hotspots, which are cellular data device that create a small WiFi hotspot that other nearby devices can use to get online, sharing in the cellular data connection. A smartphone or tablet can usually create a personal hotspot that functions like this to share its connection, or a small dedicated device called a MiFi, Mobile Hotspot, or Jetpack can do the job too.

The radios involved and technologies underlying WiFi and cellular data are different – a cellular booster is not going to help you pull in a WiFi signal from farther away, and extended-range WiFi equipment will do nothing to improve your cell phone reception.

And needless to say – cellular and WiFi radios are completely different and not compatible with TV antennas either.

Byte Sizing – What Exactly Is Data?

When you're hooked up to cable internet or DSL, data is usually unlimited and unmetered, and you can use as much as you want without worrying about paying overage charges.

When you go mobile, that changes.

When you first hit the road, you probably aren't used to really considering how much data you're using.

Most sources of mobile internet are metered, and you must be in tune with how much internet data you're using so you can determine how much you want to pay for in advance, and so that you can avoid surprise overage charges or having your connection throttled down to a slow-speed crawl.

But unlike talking on the phone, usage is not generally measured in how many minutes you're online. Only in rare places, such as on cruise ships, is data sold by the minute.

In most cases, mobile data is sold in virtual "buckets" of a certain amount of data usable over the course of a month. This data is measured by the gigabyte or megabyte.

But how much stuff does a single gigabyte equate to?

Everything encoded in digital format for storage in a computer or transmission over a network is made up of "bits" – literally, zeros and ones. Eight bits make a byte, and a byte generally represents a single character of text.

It takes 1024 bytes to make a kilobyte (KB), and 1024 kilobytes to make a megabyte (MB), and 1024 megabytes to make a gigabyte (GB).

Text takes very little space – a book like this one (stripping away images) is only a few hundred thousand characters, less than 400KB. Mathematical data compression techniques work incredibly well on textual data – making text take up even less space in practice.

In other words – the amount of text you transfer (including emails without attachments) is hardly worth worrying about.

Pictures and music, however, start to move the needle a bit.

A typical 8-megapixel digital camera takes photos that average around 2.5MB in size. In other words – when it comes to data, pictures are actually worth several thousand words!

14

The modern web has grown very graphically rich – and web pages can easily consume 1MB to 5MB per page viewed, and sometimes even more.

Streaming music online can consume 30MB to 90MB or more per hour, depending on the quality.

But if you really want to burn through data – video is a substantial data hog. Especially with high-definition (HD) video, data amounts quickly begin to be measured in "gigs."

Uploading & Downloading

Data transfers on the internet are a two-way street, and the connection is metered in both directions.

Uploading or "upstream" connections is the data that is transferred from your computer to another computer or server. This might be a small bit of data – such as submitting a search to Google or posting a status message on RVillage. Or you may be uploading a large file – such as posting a video to YouTube.

Downloading or "downstream" connections refer to data that is transferred from another computer to yours – for example, reading a blog and downloading text and images, viewing a video on Netflix or YouTube, reading your Facebook home feed, or downloading a computer operating system update.

Video and audio chat applications like FaceTime and Skype use substantial amounts of both upstream and downstream data simultaneously.

To calculate your data usage, you have to keep in mind the total consumption both ways.

What Isn't Considered Data Usage?

There are plenty of things you can do on a computer or smartphone that will not count against your monthly data usage.

If you're not currently connected, then you're obviously not consuming any internet data.

And even when you are connected, anytime you're viewing files that are already stored on your computer, tablet, or phone – you don't have to worry about data usage.

Viewing files you have stored locally doesn't use up data – only transferring them via the internet does.

Viewing photos that you copied from your camera or phone to you computer? No data usage – unless you decide to share them on Facebook, Instagram or upload them online to a service like Flickr, Picasa, or SmugMug.

Reading an ebook that you already downloaded? Usually no data usage is involved once the book has been downloaded the first time.

If you make cellular phone calls using your carrier's regular voice service, the call will not count against your data usage. However, if you use an internet-based service like Skype, Google Hangouts or FaceTime to have an audio chat or make an outgoing call, that will count against your data.

Benchmarks for Common Internet Tasks

To help you better understand how quickly various internet tasks can burn through data, we've taken some measurements of how much data typical tasks can consume.

Actual consumption can and will vary a lot - these are just some rough examples based upon a few rounds of real world testing.

Common RVer Tasks:

Plan Your Route - Using RVillage.com and campendium.com to scout out future potential campsites, and then planning a route in maps.google.com, including checking out the satellite view of the destination campground to make sure the spot looks like a good fit, and finally scoping out fuel prices along the way at gasbuddy.com: **17.6MB**

Pay Bills - Sync transactions to Quicken, check balances and schedule two credit card payments online at two different online banks, transfer funds between accounts, and check in on investments: **13.7MB**

Check in on RVillage: Visit RVillage.com, update present location, explore the map to see who else is around, and make a post to the news feed saying hello: **9.8MB**

Post a photo to Instagram: Posted a photo of us taking this measurement of data consumption of posting a photo: **2.6MB**

Send an email to Mom: A loving text email used 70KB, with a "large" picture attached: **747KB**

iMessage Chat: Short back and forth text chat, including a photo and a contact transfer: **500KB**

Tasks by Time:

To make it easier to understand how data usage adds up, we performed some everyday tasks for set periods of time and measured the data consumed. We tested five minute at a time, and multiplied by 12 to give a per-hour rate.

Your mileage will vary greatly - but this gives a good ballpark figure to start with.

As you will see - a typical 5GB (5,000MB) data plan can be used up incredibly quickly, with just an hour or two of average web surfing per day likely to use up the entire allotment within a month.

Multiply by two or three to allow for multiple users sharing a connection, and it goes even quicker.

And if you throw HD video into the mix - your monthly bucket can end up gone in no time!

Online Task	Data Used per Hour
FaceTime Audio Call	40MB/hr
Skype Audio Call	46MB/hr
Listening to Pandora (Standard Quality)	52MB/hr
Reading / Posting in RV-Dreams.com Forum	71MB/hr
Online Gaming (Typical Usage)	75MB/hr
Reading Gmail	77MB/hr
Skype Low-Res Video Call	96MB/hr
Browsing RV Blogs	137MB/hr
Browsing Facebook	140MB/hr
Actively surfing twitter.com (including skimming links)	155MB/hr
Listening to iTunes Radio	180MB/hr
Surfing Google News, reading top stories	186MB/hr
Watching YouTube (360p Resolution)	221MB/hr
Listening to Pandora (High Quality)	300MB/hr
Watching Netflix (iPad SD Playback)	384MB/hr
FaceTime HD Video Call	408MB/hr
Watching Technomadia Video Chat Archive	516MB/hr
Watching Netflix (Desktop SD Playback)	1,104MB/hr
Watching Netflix (iPad HD Playback)	1,656MB/hr
Watching YouTube (1080p Resolution)	1,920MB/hr

Rule #1 – Reset Your Expectations!

Mobile internet has come a long way since we first hit the road in 2006 (yeah, yeah...we used to surf uphill both ways on 2G 1xRTT!).

Chris has been utilizing mobile internet since 1996, years before even WiFi came along. Cherie was an avid mobile technology user as well and owned one of the very first internet-enabled smartphones way back in 2000.

The technology for connecting while on the go has advanced at an incredible pace – there is a vast difference in speeds and coverage today, and it's only getting better.

The price per GB of data has plummeted too, though typical monthly usage has gone up even faster than prices have gone down – so things certainly do not feel any cheaper!

Public WiFi spots are all over now, and there have been so many improvements in cellular coverage that you can now get usable connectivity across the bulk of the nation.

And sometimes, you can even get connection speeds while on the go that actually exceed what you could get via the fastest fixed-place connection.

> But despite all these advancements, there are still limitations and plenty of frustrations.
>
> The most important thing you can do to prepare yourself for relying on mobile internet is to reset your expectations.
>
> Be ready for the bad days.
>
> The slow days.
>
> The no days.

19

We're not trying to scare you away, but we do want to make sure your expectations are realistic. Keeping online most of the time while traveling is entirely possible – but it's not necessarily easy or cheap.

There is a scene in the film RV where the late great Robin Williams is standing on top of his rig like the Statue of Liberty, trying desperately to send an email only to have his battery die just as his dozenth attempt looks likely to complete.

It is an especially funny bit for us RVers...because many of us have been in that exact situation (and pose) way too many times.

The real secret to connectivity on the road is learning to be flexible and embracing rather than struggling against the constantly changing ebbs and flows of bandwidth.

Living as a technomad, some days you will have a connection that seems as if you are plugged directly into the heart of the internet, and other days you will be wishing for an upgrade to IP over carrier pigeon (en.wikipedia.org/wiki/IP_over_Avian_Carriers).

Navigating mobile internet is not going to be anywhere near as easy as just plugging in a cable like one you might have had in your fixed home. And sorry to say, they don't even make a cable long enough to take it with you across country (we tried, it got tangled up in our axles).

You will be battling:

- Intermittent and variable connections.

- Varying speeds – from frustratingly slow to blazing fast.

- Bandwidth cap limitations on how much data you can use.

If your mobile livelihood absolutely depends on keeping connected, you will have to carefully plan your mobile life around this need.

This could mean altering routing and carefully planning where to stop to get work done.

It may mean having to move on sooner than you're ready to search for better bandwidth, and may even mean afternoons spent at the local Starbucks, McDonald's, brewery, or library to soak up some WiFi.

If you can turn the inevitable frustration around – from "Gah! I have to go find WiFi" to "Oh darn, I have to indulge in a local craft beer while I get some work done" – you'll be better able to thrive with this lifestyle.

It may mean that some days you will find yourself actually thankful to be paying overage fees when a single source of mobile data is your only option in a specific location.

No matter how many backup plans you build into your connectivity arsenal, when you absolutely need to get online, that's inevitably when the glitches will emerge.

No matter how much time and money you invest in ensuring great connectivity, there will be times that you can't get a stable connection to do everything you need from the location you want to be at.

You need to be ready to compromise.

Mobile equipment can (and will) fail, firmware patches and upgrades can cause unintended problems, weather can interfere, your exact parking location can influence your signal, and even something as invisible as your neighbor's microwave oven can conflict with your WiFi network, knocking you offline until their popcorn is ready.

You need to realistically set your expectations, as well as the expectations of the people depending on you being online – such as clients, co-workers, family, and friends.

> There will inevitably be compromises in connectivity in exchange for your mobility.

Very seriously consider the costs of assembling your mobile internet options. Decide for yourself how much it is worth spending to try to cover your bases and how much internet access *you really need.*

Even if your income source does not depend on internet access, you may need to adjust your expectations around how much you rely on the connectivity for personal reasons – such as email, Facebooking, viewing video content, banking, bill paying, looking up information about your next destination, online learning, gaming, and keeping in touch with loved ones.

Some days, the daily tasks you depend on may prove too frustrating to deal with.

If you're used to streaming TV and movies over your fast unlimited cable internet, you may need to adjust your viewing habits to include other sources of entertainment.

If online gaming is your entertainment of choice, you may need to resign from your Warcraft clan and focus on single-player or turn-based games instead.

And if you've gotten hooked on video chatting, you may need to actually resort to old-school voice phone calls every so often.

No matter what you do, there will be days that staying connected is more of a headache than it is worth. The most important rule for staying connected on the road is that you need to be mentally prepared for these days.

Our secret? Always have a box of wine onboard.

Assembling Your Arsenal

There is currently no one single technology for keeping online that is appropriate for all the different situations mobile users might find themselves in.

If mobile internet is important to you, you will likely need to assemble a diverse collection of tools to keep you connected.

The most fundamental key to success staying online while on the road is having multiple pipelines ready to be tried at each location.

When Plan A is out of range or overloaded, Plan B suffers a hardware failure, and a tree is blocking the signal to Plan C – what will you try next? How much redundancy do you need?

Each individual solution has its pros and cons, and will be more ideal in some locations than in others. And there are no easy universal answers.

Sometimes what works best in a given location even changes based on the time of day, or the weather.

Your ideal arsenal is going to be very personalized to you and dependent upon several factors including:

- **How important is the internet to you?**

  Do you need to be online all the time? Or can you be offline for a few hours? How about a couple days or (gasp!) weeks at a time?

- **How much internet data do you need?**

  Do you just need to send a photo of today's great view to the grandkids, schedule a banking payment, update your location on RVillage, and do a little surfing? Or do you need to regularly transfer big files, VPN into remote servers, watch nonstop cat videos, trade stocks online, or attend lots of online video classes?

  How much of this is really a need versus a luxury for you?

- **How many devices do you need to keep online?**

  Keeping one laptop online is different from keeping a large multi-device family connected.

- **What is your style of travel?**

  Are you planning to mainly boondock in remote locations far away from civilization where few cell towers exist? Or are you moving between commercial RV parks primarily in urban areas? Or will you have some combination of styles, including some public campgrounds and courtesy parking with friends?

  Will you be traveling within the USA, or will you be crossing borders into Mexico, Canada, or heading farther abroad?

  How quickly will you be changing locations?

- **What plans and devices do you have now?**

  Do you have existing contracts to contend with? Do you have a coveted grandfathered-in unlimited data plan that you'd like to keep? What gadgets do you have or anticipate getting?

- **Your budget?**

  The cost of staying connected can add up quickly, between equipment and monthly fees. Can you afford to keep multiple redundant backup options activated, even if you won't be able to access them at some locations?

Free and cheap options will have trade-offs for convenience. And even expensive options come with frustrations.

If you're not hitting the road in the next couple of months, please don't jump into buying all your equipment right away.

Technology changes so quickly that you are best off leaving the assembly of your connectivity arsenal until much closer to when you will actually hit the road. We recommend no sooner than 6 months before you hit the road, but 3 months is even better.

There will be new options to consider, and what you buy now may become obsolete by the time you need it.

We personally tend to go through a total connectivity refresh at least every year just to keep up, though not everyone needs to be as focused on staying on the bleeding edge.

Our current arsenal, which we detail in the appendix of this book, may be overkill for your needs, and for some may not be robust enough. Throughout this book, we break down the trade-offs so you can decide for yourself how much of this technology you personally actually need.

Also – don't feel you have to have your total connectivity solution built on day one. Yes, some things will be easier to install when you have access to ladders and tools and geeky friends. But you can always change things up later if you find something isn't working for you or you decide to change up your style of travel.

And never forget that trade-offs will be inevitable. It may be frustrating in the moment, but the occasional bout of disconnectedness is a small price to pay for all the incredible perks of a mobile lifestyle!

Cellular Data

Cellular data is probably the easiest and most accessible option in most places across the USA. This is where some of the greatest advancements in speed, coverage, and reliability have happened over recent years.

Cellular data allows you access to the internet anywhere your devices can get a cellular signal from your carrier(s).

Each cell phone company (aka carrier) builds cell towers in locations it has customers to serve, and each tower transmits a signal of varying power that can be picked up by devices within its coverage zone, called a cell.

Cellular Internet

Quick Glance

Pros

Widely available

Easily accessible

Can be blazing fast!

Cons

Expensive

Data usage caps

Variable signals

Those individual cells may range in size from the size of a city block to the size of an entire town (or larger!), and they are designed to overlap, creating a network of coverage.

As you move out of one cell, the connection to your device is handed off to the next tower – usually fairly seamlessly. The places where there are gaps between the cells where no tower reaches are known as dead zones, and though carriers try to avoid having any, they are sometimes inevitable.

The farther you move away from the tower in the center of a cell, the worse your signal may get – resulting in dropped calls and slower data. In addition to distance, the signal can also be impacted by terrain, buildings, obstructions, and radio interference.

The quality of your connection will also be affected by how many other customers are utilizing the same tower you are connected to. A given cell can only handle so many connections at once; to increase capacity when an area gets overloaded, the carriers build more towers, breaking larger cells into multiple smaller ones.

More than likely you're already carrying a cellular-equipped mobile internet device – such as a smartphone or tablet.

But as simple as it can seem on paper, cellular is also a sometimes hopelessly confusing subject – primarily because there are just so many options! You have to choose which carrier(s) you want, which plans make sense, what equipment to purchase, and how much speed and data you actually need.

The biggest downsides to cellular data are:

- You are dependent on where your carrier has built towers and/or has roaming partnership agreements. No one has coverage everywhere, and even if you bought a plan and device from each of the major carriers, *you would still* encounter times you just can't find a signal.

- Bandwidth caps and network optimization practices are in place to limit how much data you can use per month.

  There are relatively few unlimited mobile data plans left out there that can be purchased now, and most of them have substantial limitations lurking in the fine print. A lot of the unlimited plans also tend to be on carriers with limited coverage maps and/or slower speeds. It's a trade-off – great coverage with limited caps, or crappy coverage with unlimited usage.

  Oh, and even though a plan might advertise "truly

unlimited data" – read the fine print: many times that doesn't include access for use on your computer, just on the mobile device itself.

Nationwide Carriers

The first choice to make is which carrier, or carriers, you should get service with to best cover your mobile data needs.

In the US, the four major nationwide carriers are:

- Verizon
- AT&T
- Sprint
- T-Mobile

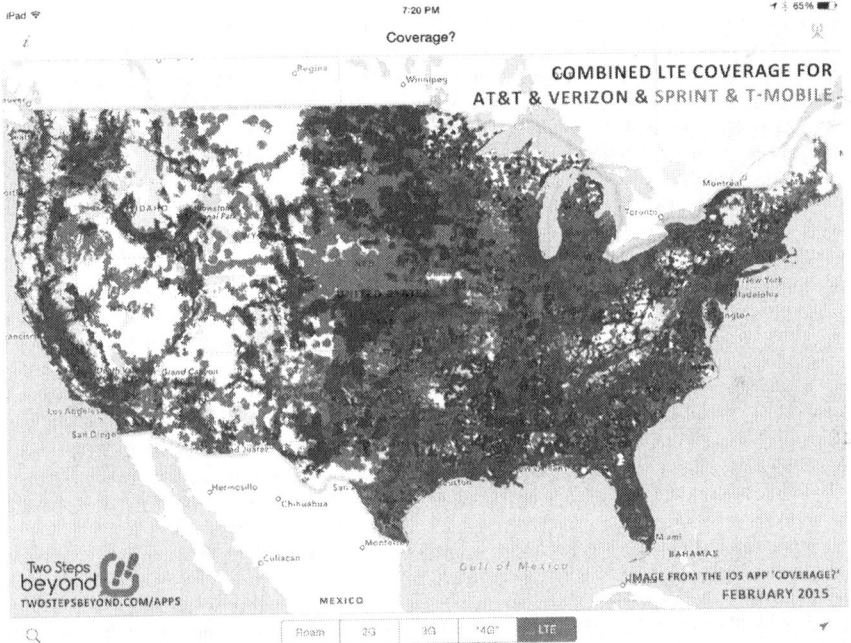

All four are converging towards the same underlying fourth-generation (4G) cellular network technology, known as LTE. But they all have very different legacy 2G and 3G networks, coverage maps, compatible devices, supported frequencies, and expansion plans going forward.

If you were living stationary in one city or neighborhood, you could ask friends for their experiences with their carriers to determine which would serve you best.

But as a traveler, you will be moving around a lot – and in different locations, different carriers excel. You need to pick a carrier that is well suited not just to your home turf, but also for all the places you plan to go.

The most important point we want to highlight here is that there is no singular best carrier for a nomadic traveler.

Verizon

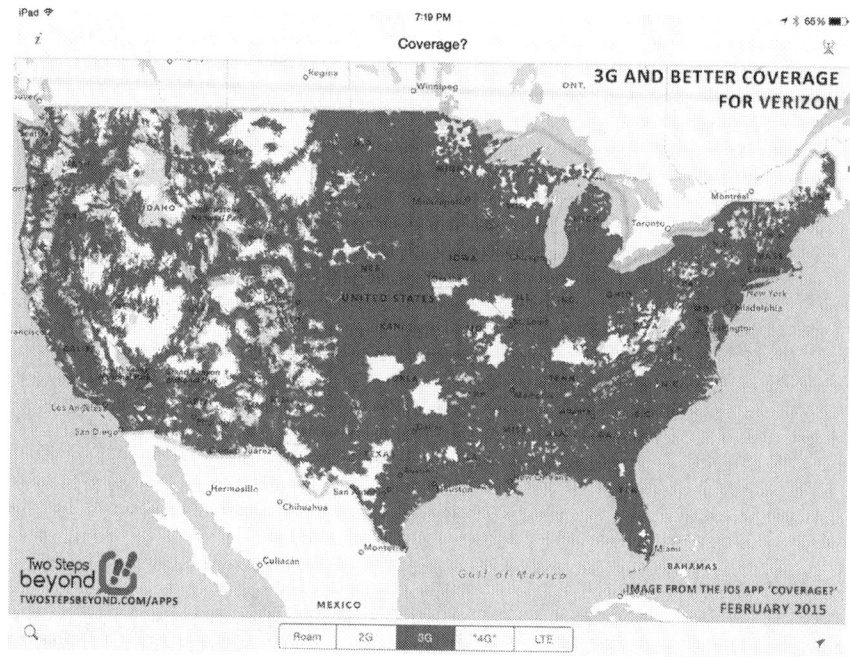

Verizon is the largest cellular carrier in the US.

Verizon has the most overall coverage, the most deployed LTE, and great overall performance. Verizon has a great future roadmap too – with an ever-expanding network and new XLTE service coming online for even more speed in key areas.

For all these reasons, if you're only going to invest in one network – Verizon is the natural top choice.

But Verizon is also pricey. And because Verizon's network is known to have the widest coverage and has become a top choice for RVers, it's actually not uncommon to pull into a large RV park and find the local Verizon tower overloaded and sluggish during peak times.

28

AT&T

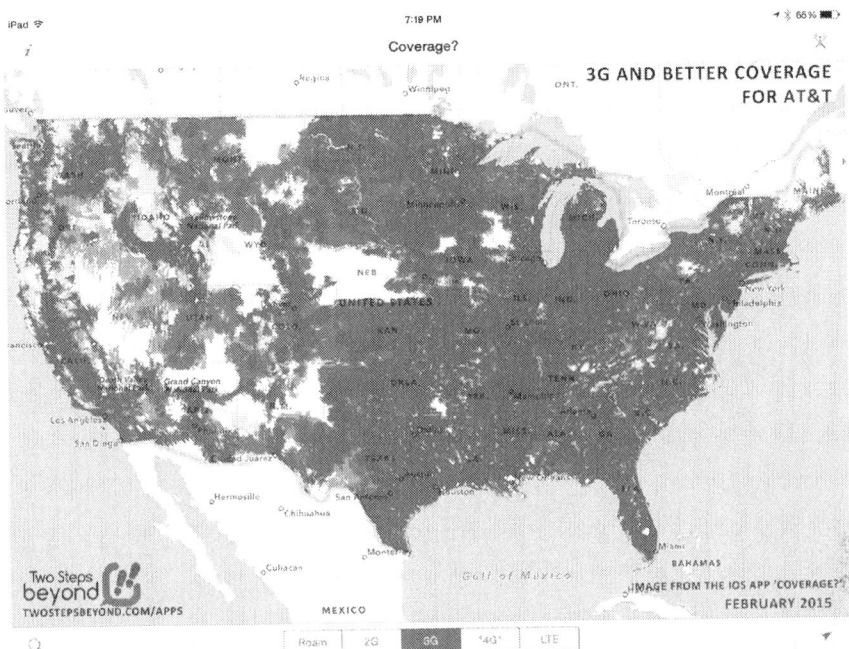

3G AND BETTER COVERAGE FOR AT&T

IMAGE FROM THE IOS APP 'COVERAGE?' FEBRUARY 2015

AT&T is the second-largest carrier and is a formidable rival and a great complement to Verizon for us nomads.

AT&T's LTE network lags Verizon in coverage and speed, but AT&T's older 3G and 4G networks have great coverage and speeds that blow Verizon's 3G out of the water.

There are plenty of locations we have been where the best Verizon can manage is 3G or very weak LTE, and AT&T ends up delivering a much better overall experience.

New in 2015, AT&T is now offering 'rollover data' on its mobile share data plans - if you don't use up all of your data one month, it rolls over to be used the next month.

By keeping both Verizon and AT&T connections on board, we can play to their strengths – and go with whichever has the best connection and fastest speeds in a given location.

Sprint

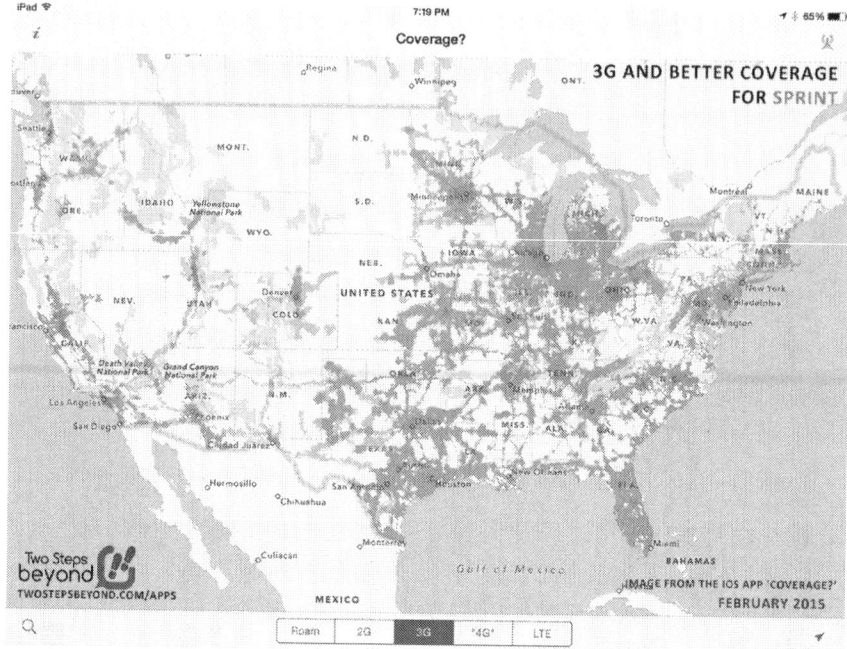

Sprint's biggest advantage is that it still offers unlimited on-device data plans, though Sprint has now begun to throttle heavy users to slower speeds, diminishing this perk.

The biggest downside of Sprint for nomads is the limited coverage map. The vast bulk of Sprint's usable fast data coverage is pretty much only in core urban areas and along major interstates. Anything outside of that, and you're roaming with very slow speeds – if you can get online at all.

And despite offering unlimited usage on Sprint's core network, when you roam off of it you are facing very limited roaming data caps that could leave you either suddenly cut off or with a surprise overage bill.

However, if you're planning to stick to urban areas, Sprint might be worthwhile. Sprint has begun rolling out LTE rapidly now, and the new tri-band LTE Sprint Spark network shows great promise for the future.

Sprint is also aggressively pursuing roaming agreements in rural markets to partner on bringing LTE speeds to more places.

Just be sure to avoid counting on Sprint's orphaned and limited WiMAX 4G network, which will be shut down by the end of 2015.

T-Mobile

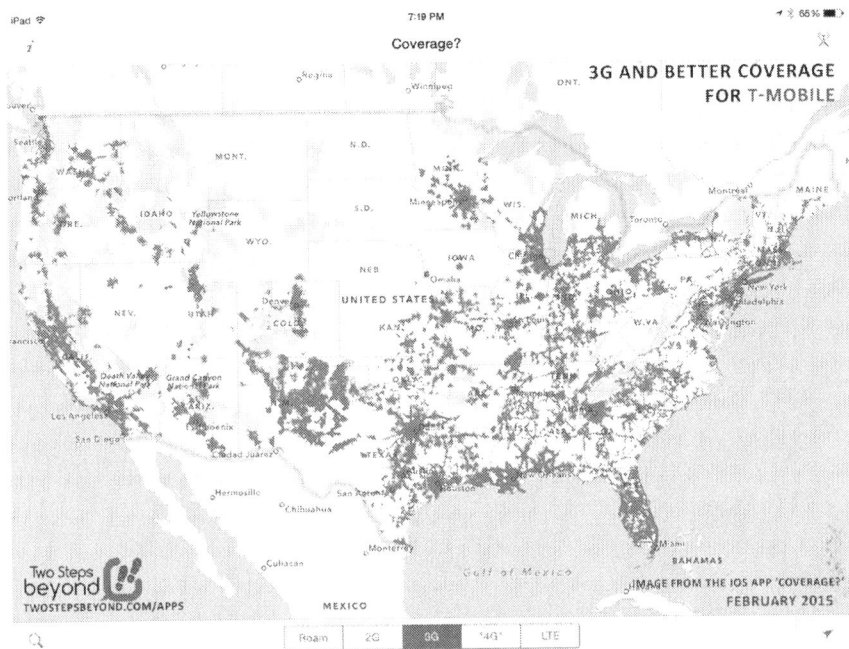

T-Mobile is an interesting contender that has made several moves to shake up the traditional rules of the mobile industry by being the unconventional "uncarrier." T-Mobile has challenged the other major players to move away from contracts, subsidized devices, and towards lower prices overall – and the pressure from little T-Mobile has changed the entire industry for the better.

Unique features T-Mobile offers include unlimited free music streaming, and unlimited free (but slow) international data roaming.

T-Mobile also offers 200MB/mo of "Free Data For Life" on compatible tablets - with no contract or ongoing billing of any sort required.

(For details how to get this: http://www.rvmobileinternet.com/resources/scoring-free-data-for-life-from-t-mobile/)

T-Mobile's latest move (January 2015) has been introducing a 'Data Stash' rollover data feature to most T-Mobile plans, which allows you to bank unused data which then does not expire for an entire year.

T-Mobile has an exceptionally fast 4G network and has also begun aggressively rolling out LTE. T-Mobile also offers an unlimited on-device data plan that might be very attractive for high bandwidth users.

T-Mobile's Achilles heel, however, is coverage: away from urban areas, T-Mobile's signal is usually nonexistent – making T-Mobile a tough choice for travelers.

31

Keep Current Alert:

The carriers are constantly changing up their offerings and features to stay competitive with each other.

Before choosing your carriers, be sure to check the RV Mobile Internet Resource Center for our current take on each. We keep this article updated with our analysis of each carrier as announcements are made:

The Four Major US Carriers - Which is Best for RVers?

(http://www.rvmobileinternet.com/resources/the-four-major-us-carriers-data-comparison/)

Which to Choose?

For nomads who need the widest amount of cellular coverage, we generally find that combining Verizon and AT&T gives the most options. There are simply locations where one carrier excels over the other.

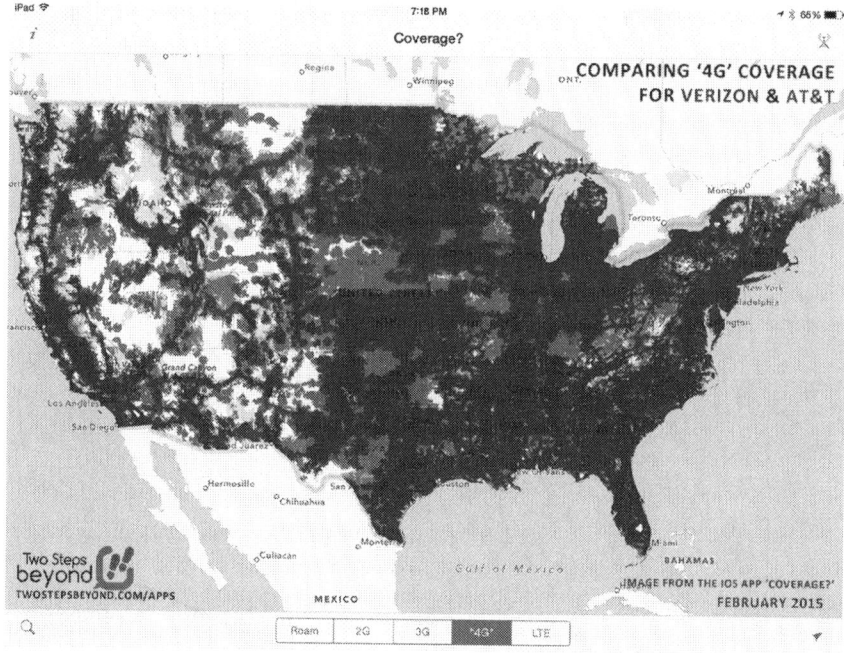

For nomads seeking to have only one carrier, Verizon is usually a solid choice – but AT&T's coverage map should also be considered.

And of course, even having both is no guarantee that you'll be able to get online everywhere.

Each carrier offers plans directly that you can obtain either on their website, in stores around the country, or via third-party resellers. You can get plans that are for data only, or for data combined with smartphones or tablet

32

service. We'll cover some considerations for selecting a cellular plan direct with the carriers later in the book.

Regional & Local Carriers

In addition to the big four national carriers, there are a number of smaller regional and even local carriers that own and operate their own wireless networks.

Some of the larger examples of this sort include U.S. Cellular, C Spire Wireless, nTelos, Cellcom, and Cellular One.

These smaller regional carriers are usually poor choices for travelers, unless you know that you are primarily going to be spending time in areas where they have a strong presence.

Even if the carrier has nationwide coverage through roaming agreements, if you're utilizing the service primarily outside its home region, you can find yourself running into all sorts of limitations imposed that can even include having your service summarily terminated.

Knowing Where to Find Signal

Part of selecting the right carrier(s) for you is knowing where they have the coverage you desire. And once you have them as part of your mobile internet arsenal, it's helpful to know where along your routes you'll get signal.

Here are some handy resources for tracking this sort of information down:

Checking the Carriers' Online Maps

- Verizon – www.verizonwireless.com/wcms/consumer/4g-lte.html
- AT&T – www.att.com/maps/wireless-coverage.html
- Sprint – coverage.sprint.com
- T-mobile – www.t-mobile.com/coverage.html

Coverage? App

Although we can go to each carrier's maps online to scout out ahead if our next campground will have coverage for our carriers, we decided to make it even easier. We wrote an app for that!

Coverage? (available currently for iOS, and

33

hopefully soon for Android – www.twostepsbeyond.com/apps/coverage) overlays our versions of the four major carriers maps, so you can create a personalized coverage map for the carriers you travel with.

The maps are stored on device, so you don't even need to have coverage right now to find out which direction to head so you can participate in that 2 p.m. webinar.

All of the coverage maps displayed earlier in this chapter are from our app.

Crowdsourced Coverage Maps

Of course, just because a carrier claims they have coverage, doesn't mean you'll be able to find it. There are some wonderful resources out there that aggregate crowdsourced signal and speed reports to create a coverage map based on measured reality.

Here are some of our favorites that we utilize:

- **OpenSignal** – Online at www.opensignal.com and they also have an iOS and Android app of the same name.

- **RootMetrics** – Online at www.rootmetrics.com and they also have an iOS and Android app called CoverageMap.

- **Sensorly** – Online at www.sensorly.com and they also have an iOS and Android app of the same name.

The downside of crowdsourcing is the maps available are only as useful as the data they collect from users of their apps. These resources tend to have good data for urban areas where they have a strong user base. But when you get to smaller cities and back roads, these maps commonly show no coverage at all.

These apps are a great compliment to Coverage? – first you can research where the carriers claim to have service, and then you can research user-submitted reports to get an idea on actual performance.

MVNOs & Prepaid Plans

Mobile Virtual Network Operators (aka MVNOs) do not own their own cellular networks, but rather they buy service in bulk from the major carriers and resell it under their own brands with their own terms.

Via an MVNO you can get better deals with no contracts and higher bandwidth caps than you ever could by buying direct.

Usually an MVNO is not allowed to advertise who is providing the underlying network, but if you explicitly ask or do some online digging, it isn't hard to figure out which data network is behind any given device or plan.

Many MVNOs are actually even owned and operated by the big carriers themselves – allowing them to target smaller markets and niche services without diluting their national brand.

Most carriers also offer prepaid plans that allow customers to only pay for service when they need it, rather than month to month. This makes setting up service super-easy without requiring signing contracts or going through credit checks.

Some of the more popular mobile internet MVNOs and prepaid data only services include:

Verizon Based

- **Verizon Prepaid** (www.verizonwireless.com/prepaid/prepaid-jetpack/) – Available direct from Verizon on its LTE network.

  - Offers prepaid Jetpack hotspot plans starting at $15/week for 250MB, $60/month for 3GB, and up to $90/month for 10GB of LTE data.

  - Purchase of a Jetpack hotspot device for around $75 required.

  - No additional activation or reconnection fees; use it as you need it with no contract or credit checks.

- **StraightTalk Hotspot** (www.straighttalk.com) – Available at Walmart and online, this hotspot plan is on the Verizon's network and requires a hotspot purchase of around $80.

  - Prices are on a pay-as-you-need-it basis and can be purchased in smaller increments with usage spread

35

out over up to 2 months – starting at $15 for 1GB, up to $75 for 7GB. Ideal for occasional use, pricey for higher bandwidth needs.

- Speeds are reported to be slower than on compatible plans direct with the carrier.

- Plans purchased at Walmart are now on Verizon's LTE network, however plans/devices purchased online are still offering only 3G.

- StraightTalk also offers a similar LTE service on AT&T's network - so make sure you know what you are getting.

- **Blue Mountain Internet** (bmi.net/internet/mobile-broadband-rental.html)

  - BMI offers no-contract 3G/4G Verizon-based plans starting at $39.99/month for 1GB up to $99.99/month for 10GB of data – providing a range of options depending on your bandwidth needs. There are quarterly discounts as well.

  - They claim that they provide "optimizer software" for Windows & Mac machines that can effectively triple your usage for non video files.

AT&T Based

- **GoPhone Mobile Hotspot** (www.att.com/gophone) – AT&T directly offers its own prepaid service, which is available at Walmarts around the country.

  - You purchase a hotspot MiFi device for about $100, and then you fund the account as needed by purchasing refill cards.

  - The costs range from $15 per week for 250MB up to $50/month for 5GB of 4G data.

  - The gotchas to watch out for include: Service is only available on AT&T's owned network – none of its roaming areas – and your refill card balance will expire anywhere from 30 days to a year after you redeem part of it.

- **StraightTalk Hotspot** (www.straighttalk.com) – Available at Walmart only, this hotspot plan is on AT&T's network and requires a hotspot purchase of around $80.

36

- Prices are on a pay-as-you-need-it basis and can be purchased in smaller increments and usage spread out over up to 2 months – starting at $15 for 1GB, up to $75 for 7GB. Ideal for occasional use, pricey for higher bandwidth needs.

- Speeds are reported to be slower than on compatible plans direct with the carrier.

- Plans purchased at Walmart are now on AT&T's LTE network, the plan is not available online.

- StraightTalk also offers a similar service on Verizon's network - so make sure you know what you are getting.

Sprint Based

Sprint is a favorite network provider of MVNOs, so there are a lot of choices out there.

- **Karma Mobility** (www.yourkarma.com) – Karma is perhaps the most unique MVNO out there. Its only product is a stylish $149 personal hotspot, and you can then buy data (which never expires) for $14/1GB, or $99/10GB – with no contracts, monthly fees, or fine print.

  The Karma hotspot is designed to be shared; if another user connects through your Karma hotspot, you both earn a bonus 100MB of data, and their usage does not count against your data pool.

  The original Karma device was on Sprint's soon-orphaned WiMAX network, but it is being replaced by a new Karma Go model designed to run on Sprint's Spark tri-band LTE network. The current estimated ship date for the Karma Go is April, 2015.

- **Virgin Mobile** (www.virginmobileusa.com) – Virgin Mobile, owned by Sprint, offers no-contract Broadband2Go plans with scalable data rates. They have monthly plans starting at $25/month for 1.5GB and daily plans at $5/day for 250MB.

- **EVDODepotUSA** (www.evdodepotusa.com) – Aimed at rural customers, they offer an unlimited Advanced Surfer plan on Sprint's 3G network for $119/month with a 6-month contract, or a 60GB/mo LTE plan for $139/

37

month. (Heads up: After 6 weeks of communication, this company failed to actually ship us a test unit they repeatedly promised.)

- **Wireless n WiFi** (www.wirelessnwifi.com/High-Limit-4G3G-Service) – Offers on a limited basis a high-limit plan on Sprint's 3G and LTE networks for $89.99/month that provides 60GB per month of usage.

- **FreedomPop** (www.freedompop.com) – Offers a free 1GB of data per month, or plans ranging up to 10GB a month for $21.99/month. Great if you just want Sprint as a backup option for a little extra bandwidth.

- **TruConnect** (www.truconnect.com) – TruConnect offers a pay-as-you-go mobile internet plan on Sprint's network. You pay a $4.99/month access fee and then 3.9 cents per MB you use. If you just want to have Sprint on board in case other options fail, that's a pretty affordable way to get a little bit of bandwidth redundancy. Just be aware that if you need a lot of data, this equates to nearly $40 per GB.

Any of these plans could be useful as a backup to have on board if you find yourself in a Sprint area with a lack of other options.

Bandwidth-hungry nomads can keep a Sprint high limit or unlimited plan around for bulk downloads when they are in a covered area.

There are so many more choices out there; for an overwhelming list of the options, check this list of MVNO's on Wikipedia (en.wikipedia.org/wiki/List_of_United_States_mobile_virtual_network_operators).

The Need for Speed

Cellular companies love to brag about how fast their networks are – but once the connection is fast enough to stream HD video (around 5Mbps – megabits per second), do most mobile users really need any faster?

Faster and more responsive surfing is nice – but with monthly data caps and potential overage charges, there are very real downsides to speeding down the information highway too quickly.

Often, we find ourselves wishing the network were actually slower so that we wouldn't accidentally burn through data so fast!

Why, then, are carriers so gung-ho about fast networks?

The key is capacity.

The faster the network is able to serve you whatever it is you've asked for, the faster it is able to get on to serving the next person. With only so much spectrum to go around and networks in many areas already oversaturated, more speed is almost a matter of survival.

Slower 2G and 3G data actually hog up more network airtime – costing carriers more to serve than sending the same data faster to LTE users. Once the novelty of LTE wears off, expect carriers to make a big push to get users off the 3G networks and onto LTE as quickly as they can.

Verizon has already expressed intentions to begin bringing out LTE-only devices in 2015, which will drop support for the old 3G and 2G networks entirely. And unlike earlier mobile hotspots, some newer devices even lack the ability to force 3G-only mode when there is an LTE signal to be found.

Cellular Pitfall: Data Caps

Cellular data is generally dished out in monthly buckets with data caps (also called bandwidth caps), setting a limit for how much data you can use before you are throttled down to a slower speed or charged extra in overage fees.

The monthly cap on a plan might be marketed as 5GB, which allows you 5GB of data, totaling up your usage both sending and receiving.

Things to keep in mind with cellular data caps include:

- Even if the file you're transmitting doesn't complete, you are charged for the data used. So if you're uploading a video to YouTube and it fails halfway through – you've still used the data to load half the video, even though your mission was not completed.

- If you don't use your full data cap by the end of your billing month, on Verizon & Sprint you lose it. AT&T and T-Mobile recently started offering rollover data, but on AT&T in particular there are substantial limitations on how this works in practice.

- The carriers oftentimes don't tally up your data usage in real time – it can be hours behind. This can make it very difficult to manage your data as you get close to your cap.

- If you exceed your cap, your carrier may charge you overage fees – which are usually billed in increments of 1GB – typically $10–20/ GB on most current plans. But overage charges can be painfully

39

more expensive on older plans or if you are caught roaming. And even if you only go slightly over by just a few KB in a month, you're charged for a whole extra GB.

- Because so many users get upset with unexpected overage fees, some plans now offer unlimited usage with a high-speed data cap instead. When you exceed your high-speed cap for the month, rather than being charged an overage, your connection is throttled down to a snail's pace until the end of the month.

- Unlike when the carriers used to charge for cellular phone minutes on a peak versus nonpeak time basis, there's no equivalent with cellular data. No matter what time of day you use it, cellular data counts the same.

Sharing: On-Device Data vs. Hotspot / Tethering Data

In your research, you'll probably encounter a cellular plan that says there's included data –potentially even unlimited data. Just be sure to get clarification if the plan is for on-device-only data, or if that data is able to be shared with other devices such as your laptops.

Many phones and tablets today, especially smartphones, can be used as modems for your computers – allowing you to share your phone's data plan with your other devices. Here are the two ways:

- **Tethering** – This is the term used when you are using a USB cable to connect your cellular device to your router or computer. A nice bonus – this keeps things charged up while sharing!

- **Personal Hotspot** – This refers to when you turn your cellular device into a WiFi router creating a hotspot, allowing it to share its internet connection with other nearby devices and computers.

Many cellular plans include data for use on your phone or tablet only – such as checking email and browsing from the device. These plans prohibit sharing any of your data with other devices.

Some other plans may have the option to share data – but only if you purchase a certain amount of data first or pay an extra fee.

And even plans that support sharing data may limit the amount of data that can be used in this way, especially if you are in a roaming area.

There may be hacks and work-arounds on some devices that enable you to bypass blocks and enable sharing data without paying your carrier extra for the feature, but doing so can violate your terms of service.

Know what the potential risks are, and decide if you're willing to take them, before pursuing these work-arounds.

Unlimited Data Plans

Radio spectrum (the number of channels available to broadcast on) is a limited commodity, and only a certain amount of traffic is physically possible on any given cell tower.

And yet, the appetite for consuming content online seems to be insatiable.

To try and keep usage under control and the profits flowing in, all the cellular networks have policies in place to limit usage. Traditionally they do this by charging per megabyte or gigabyte – but to make sure that they still get to charge even the lightest users, they force you to buy monthly buckets of data to cover your typical usage, and then you pay overage charges if you go over.

Sprint and T-Mobile are the only major carriers currently offering unlimited data plans. But these plans are for data used on smartphones and tablets only, and limits (typically 5GB usage) come into place when sharing with other devices.

If you want more data for sharing with your tablet or laptop, you are required to get a more traditional plan with a limited data bucket and stiff throttling if you go over it.

Sprint, in particular, reserves the right to throttle down speeds to just 1Mbps when streaming video on these advertised unlimited plans, promising a potentially very choppy experience.

Grandfathered Unlimited Plans

AT&T and Verizon used to also offer unlimited plans, but now both have switched to selling data only in preset monthly amounts.

New accounts can no longer sign up for these plans, but some of these old plans are still grandfathered in for existing account holders in good standing who have not given them up.

If you have one of these plans, it may be worthwhile holding on to it. You might even have a valuable asset that you can sell if you no longer need it.

Verizon Unlimited Plans

Verizon continues, for now, to honor its old unlimited plans. The catch is, if you ever upgrade your phone, you will be forced to give up the old plan, unless you buy the new phone at full retail price or activate a used out-of-contract phone.

41

Many folks find this limitation to be a very worthwhile trade-off.

By default, these plans do not include tethering or hotspot support, but many folks have gotten around this by either adding on a data-sharing plan to their account with Verizon (usually around $30/month), by installing third-party apps like PDANet+, or by transferring their SIM card to a Jetpack/MiFi mobile hotspot device.

Currently, Verizon only throttles back high-bandwidth users in congested markets who are on their older 3G unlimited plans. Which means if the tower you are using is getting overall high usage and you're found to be in the top 5–10% of usage on that tower – they may cut back the speed you receive substantially until the load on the tower drops. By Verizon's estimates, this could impact customers using as little as under 5GB a month of data.

In late September 2014, Verizon canceled its announced plans to begin similarly aggressive 'network optimization' for its LTE unlimited customers, which briefly lead to a spike in sales of these plans on the secondary market.

However, in November 2014 Verizon made a policy change to no longer allow others to assume liability for these plans - meaning you can longer shop for these plans on eBay and take them over.

If you do have an unlimited plan on Verizon, make sure that you don't accidentally lose it - you will probably never be able to get it back if you do.

Potential resources:

- Ryan Maharg works for a Verizon Authorized Reseller. His specialty is the RVing market, and he claims to be able to help customers with unlimited plans keep their plans while upgrading phones. Ryan offered to share his contact information with our readers. Give him a call at 740-403-8414 or write him at ryan@wirelesssales.org and see if he can help.

- This thread on Howard Forums (www.howardforums.com/showthread.php/1814923-The-New-Improved-Update-Guide-on-HOW-TO-KEEP-YOUR-UNLIMITED-DATA) is a pretty thorough guide to keeping your Verizon unlimited plan.

AT&T Unlimited Plans

iPad: For a brief period in 2010 right after the very first iPad came out, AT&T offered an unlimited iPad data plan for just $29.99 per month.

Despite AT&T quickly realizing how much data an iPad can consume and discontinuing the plan, those who kept the plan active continue to receive this perk. And it's been transferrable to newer iPads, including LTE-enabled devices.

Unfortunately, this plan does not include personal hotspot or tethering support - however many do choose to either jailbreak their iPad, or put the SIM in a mobile hotspot with success. The plan has thus far been completely unthrottled no matter how much data you use, and the iPad can use an HDMI cable to connect into a TV set – making it an ideal media-streaming device for Netflix, Hulu, YouTube, HBO Go, and various network-specific apps. It can also be used for FaceTime, Skype, Google Hangouts, webinars and live streamed video chats.

iPhone: Unlike the unthrottled iPad plan, AT&T's grandfathered unlimited iPhone plans are now throttled in speed after around 3GB of usage a month, and even the first 3GB of data cannot be used for sharing.

With these limitations, most customers would be better off switching to one of AT&T's new Mobile Share Plans. Doing so usually saves money, increases speeds, and enables data sharing via the iPhone's Personal Hotspot feature.

There is currently a lawsuit underway regarding this practice - if this involves you, keep your eyes open for the ruling on it.

Getting an Unlimited Grandfathered Plan

Currently, only the AT&T iPad plans can be easily transferred to a new party.

Prices on eBay to take over one of these plan (not including an iPad device to run it on) range from $600-$1500 at times.

You will also stumble upon sellers who are willing to lease you their plan, which means you just pay a monthly fee to use the plan without assuming ownership of the account.

Verizon no longer allows assumption of liability of their unlimited plans.

There were however some enterprising individuals who bought up unlimited plans before the policy change, who now rent them out for a higher monthly fee. This practice is technically against Verizon's terms of service - but the risk is to the account holder, not the consumer renting the plan. Should Verizon crack down on the account holder, the renter just risks having their internet suddenly cut off. You can plans like this listed on places like eBay - shop wisely.

Be aware that there is always a risk that the carriers will start throttling or even not honoring these old plans at some point in the future. So weigh that risk into the cost you're willing to pay to take over such a plan.

Understanding Roaming

Roaming is when a cellular carrier has agreements with other networks to utilize their towers, helping the carrier provide connectivity to their customers who are just passing through areas they don't directly service themselves.

Roaming is essential to the regional carriers and the smaller national carriers, since they lack the vast networks of AT&T and Verizon. But even AT&T and Verizon use roaming behind the scenes to help flesh out their coverage maps.

The guest carriers pay very high fees for roaming data. T-Mobile has revealed that in 2014 it expects to have paid AT&T on average $181/GB for roaming onto AT&T towers!

Though carriers rarely charge you extra for domestic roaming any more – they tend to have special data roaming limits to keep your usage from costing them too much.

Here are the data roaming limits for each of the four major US carriers:

- **Verizon** – None. They truly do have the most extensive network, and actually have very few roaming partners. They impose no restrictions.

- **AT&T** – After you've used just 24MB of roaming data in a month, AT&T reserves the option to suspend your roaming services for the remainder of the month or even impose fees. AT&T is not consistent on how aggressively they enforce this, and it seems that it really depends on what network you are roaming onto and what their agreements are.

- **Sprint** – Sprint is the clearest of all the carriers – they have a 300MB roaming cap per month on all of their data plans. If you go over 300MB, they start charging overage fees of 25 cents per MB. That can add up!

- **T-Mobile** – Their caps vary depending upon how many GBs of data your plan includes. It ranges from 10MB to 200MB before they suspend your roaming service, cutting you off for the remainder of the month.

Most of the carriers will send you a text message and/or email to let you know you've hit your cap.

Right next to where how many bars or dots of signal strength you are getting is listed the carrier's name or designation. Usually this is set to your provider. When you're roaming, you might see another carrier's name listed.

By default, some carriers and devices don't switch the carrier's name displayed until *after* you've hit your roaming cap. If you're planning to be traveling a lot, you may want to call your carrier and ask if they can make that switch now.

We actually were inspired to create our Coverage? app after running into this, so we had a way to quickly check if we were in a roaming area before paying campground fees.

Roaming Tip: Beware of International Borders!

All carriers charge an arm and a leg for international roaming (including onboard cell networks offered on cruise ships). If you are going to be anywhere close to an international border, make sure to turn OFF data roaming on all of your devices.

Otherwise, you might find that you stumbled into a multi-thousand-dollar bill!

Cellular Data Gear

Once you decide on the carrier(s) you want in your arsenal, you have to decide what specific equipment makes the most sense for getting online.

The basic options include devices that are restricted to data only, or putting cellular connected tablets and phones to work serving double duty by providing an internet connection for your computers as well.

All of the big four national carriers have agreed on the same 4G networking standard – LTE. But despite the common core standard, it is normal for cellular phones and devices to be optimized to work with just a single carrier.

This is because all four carriers are building their networks on different frequency bands, and they also have older incompatible legacy networks to support 2G, 3G, and voice services.

The two most common data-only devices on the market today are USB sticks that can plug directly into your computer or router, or mobile hotspot devices with integrated WiFi routers inside that can create their own hotspot out of a cellular connection.

And most smartphones and tablets on the market today can perform a similar function, sharing their connection to become internet hotspots for your other devices.

What makes the most sense for you depends on how many devices you're trying to keep connected, how many people you need to provide internet for, how many data plans you want to keep active (each device needs its own plan), where you think you'll need internet access, and what devices you already own.

47

Cellular Modems

A USB stick device (or older style Express Card) needs to plug into something in order to be functional – either your computer or a compatible cellular-aware router. If you're traveling solo and just need to keep a laptop online, this may be an ideal solution on its own. Or if you plan to keep it plugged into a router most of the time anyway, it makes for a very elegant solution.

Because of this simplicity, dedicated modems tend to be more reliable than mobile hotspots. There is simply less that can go wrong.

Only certain routers support connecting via USB cellular modems. WiFiRanger, Cradlepoint, and Pepwave are the brands to look for with the broadest range of support.

Mobile Hotspots & Cellular Routers

- **Mobile Hotspots** (also sometimes branded as Jetpack or MiFi) are small self-contained wireless routers that receive a cellular data signal and then broadcast a WiFi hotspot that enables your devices to get online.

  Most mobile hotspots tend to be able to serve 5–10 devices at once. They usually have a battery built into the device, which allows you to take it with you. This could be useful if you want to take your laptop and tablet on a hike, on a car ride, or roaming around a city with a backpack. You can plop your laptop down on a picnic table in the park and have an online picnic with your MiFi.

  When working optimally, these can be a fairly easy plug-and-play solution ideal for users who don't want to have to learn basic networking in order to manage other more complex options.

  The downsides of this style of device is that they have a good amount of complex software installed inside them to allow them to function as a router and create a hotspot, and in our experience some of the consumer-focused mobile hotspot devices have been frustrating and buggy, especially under demanding use. Check reviews before you buy one!

- **Cellular Integrated Home Router**. While MiFis are designed to be pocket sized and battery powered, there are also heftier routers available designed for home use.

  One example is the Verizon LTE Broadband Router with Voice, which is typically marketed as a solution for providing both home internet and phone service. But some RVers have also put it to good use. It gives a home phone line that works with regular cordless sets so you can have a front and a bedroom handset, as well as a WiFi network and even wired ethernet ports, all provided via a data connection through Verizon's LTE network.

- **Commercial-Grade Cellular Integrated Router**. If your goal is building a bulletproof mobile office, rather than cobbling together pocket-sized personal hotspots or repurposing home routers, you can instead invest in a commercial-grade router with a closely integrated cellular modem paired directly to it.

  We've heard good things about the Cradlepoint ARC MBR1400, which costs $699 and comes with a built-in Verizon LTE, AT&T LTE, or Sprint LTE modem. You can still use a USB data card to provide a backup connection via another carrier too.

  This is certainly a pricier alternative – but for some mobile professionals, the increased reliability may be an essential business expense.

  Other higher-end commercial-grade products from Cradlepoint and Pepwave go even further in capabilities, focusing on both speed and reliability. For example, the $399 Pepwave Max On-the-Go router can actually load-balance your usage across five different connections simultaneously!

Smartphone / Tablet Connection Sharing

Using your smartphone or tablet to hotspot/tether as a connection point is also a very viable option. If your cellular plan allows it, you can use your data to get your computer and other devices online.

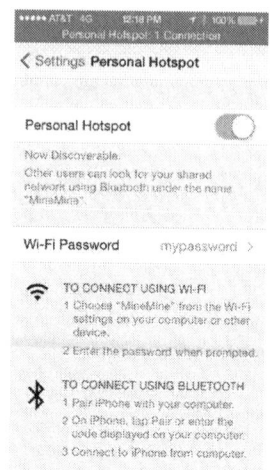

- Hotspot is when you use the device as a WiFi hotspot wirelessly.

- Tethering is when you connect directly to it with a USB cable.

Most smartphones and tablets can create a WiFi hotspot and/or be directly tethered.
Not all cellular plans include this feature, however – especially older plans, prepaid accounts, or reseller purchased plans. To get other devices online, you'll either need to add tethering onto your plan (paying $15–$30 extra per month for the privilege), or be willing to install unauthorized apps that enable the feature. Just be aware that the carriers may crack down on tethering data if you're not officially subscribed.

Each device will be different in how you turn this feature on, but for many, it's just a setting you turn on in the device's settings called Personal Hotspot – and you can easily configure a network name and password to protect the connection.

Many of the "share everything"–style plans that the major carriers are offering these days do include hotspot features and tethering with the price, but of course these plans come with data caps and overage charges.

The downsides to relying on your smartphone as a primary mobile internet connection for other devices include:

- Not ideal for multiple people in a household – What happens if the person with the hotspot-enabled smartphone takes it with him or her to run errands?

- Talking on your phone can sometimes take your devices offline or greatly reduce the network data-connection speed. So if you need to regularly talk on your phone AND be online, this may not be an ideal primary solution. (We cover this later in this chapter.)

50

- Many devices go to sleep when there's no active connection going on, so you may need to wake your device up after a period of inactivity online. Sometimes it's as simple as clicking the on button, and sometimes you may need fiddle with hotspot settings and/or temporarily engaging airplane mode to bring the connection back.

SIM Cards & Migrating Service

All LTE-compatible phones have a tiny removable sliver of plastic called a SIM – short for Subscriber Identity Module. This little chip identifies your account to the network.

This card is often found lurking underneath the battery of many phones or MiFi devices and can be carefully slid out with a fingernail. For devices that lack removable batteries, the SIM is usually in a tiny ejectable tray that can be removed by pushing a pin carefully into a pinhole.

If you move the SIM to another compatible unlocked phone or device, you are essentially transferring your service (including your phone number) to that new device.

This can be extremely handy – you can take a SIM from a Verizon Mobile Hotspot and put it into an iPad, for example – getting data on a tablet that you can still share via the iPad's Personal Hotspot feature.

Or...you can keep an old beater phone around for when you are heading out into rough conditions – such as a backcountry hike or kayaking. Before you head out, just pop your SIM into it and you can still keep online and get phone calls, without putting your flagship expensive smartphone at risk.

Or...when you are traveling internationally, rather than paying expensive international roaming fees, you can instead get a SIM (and a local phone number) from a local cellular company, and then get online much, much cheaper.

51

SIMs have been getting smaller over the years to keep pace with ever-shrinking cell phones – the original SIM from the early 1990s was the size of a credit card. But the ones you are most likely to see today are the Mini-SIM, Micro-SIM, and Nano-SIM. They are all electrically identical – so it is actually possible to cut down a Mini-SIM to put it into a device that has a Nano-SIM slot. And you can use a small plastic cradle to put a smaller SIM into Mini or Micro slot.

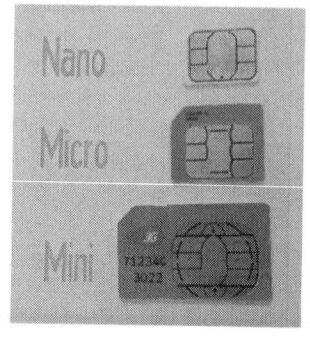

You can buy a SIM cutter tool to make the process easy.

SIM cards have always been a part of the GSM phone standard and have long been used by AT&T and T-Mobile and other GSM carriers. And SIM cards are a mandatory part of the LTE technology standard that all carriers are now using. But older 3G and earlier devices from Verizon and Sprint do not use SIM cards. If you have one of these earlier devices and want to move your service to another device, you have no choice but to call customer service and ask them to migrate it behind the scenes for you.

Locked & Unlocked Cellular Devices

SIM cards make it easy to move service between devices. If your phone is compatible with the underlying frequencies and cellular standard, by swapping SIM cards you can even change carriers. But only if your phone is unlocked.

Carriers very commonly lock new phones and hotspots sold under contract so that they will only work on that carrier's own network. If you put in a SIM from a competitor, it will just not work.

Verizon is the only major carrier that currently has a policy of selling all phones and tablets unlocked - a very nice perk.

On other carriers, you can pay full price and buy all your gadgets explicitly unlocked to start with, or there are services out there that will unlock locked phones for a fee.

Or – once your contract is expired or your phone financing is paid off, if you have an account in good standing, you can call your carrier and they will authorize unlocking your phone.

Handle getting unlocked BEFORE you head out on an international trip or try to gift your old phone to a friend on a different carrier.

52

Simultaneous Voice & Data

LTE was designed from the ground up as a data network – and it actually does not have any built-in support for traditional voice phone calls. This means that when a phone call is underway, you actually drop off the network!

This makes it frustrating to look up details or maps while on the phone making plans, and it can be especially frustrating if you are using your phone as a personal hotspot – chatting with a friend means that everyone else sharing your network is offline and twiddling thumbs!

AT&T's and T-Mobile's 3G networks supports simultaneous voice and data – so on these carriers, this has never been a major issue. Network speeds drop down to 3G while a voice call is underway, but at least the internet connection stays up.

On Verizon and Sprint, on the other hand, the legacy voice network is completely separate from the data network, so while a voice call is underway all data connections cease.

Almost all newer Android smartphones haves gotten around this problem by actually building in an entirely separate second radio for voice calls so that the data connection can remain online while a voice call is underway, but Apple never took this step with any iPhone models.

Why implement a temporary fix…when Voice over LTE (aka VoLTE) is emerging?

VoLTE is a technology that treats voice phone calls as just another data connection, meaning that the LTE network does not need to disconnect or switch into a special voice-only mode. VoLTE also opens the door to higher quality voice calls too, and to seamless switching between video and audio calls. It's now all just data.

All the major carriers have plans to roll out VoLTE, and most of them have begun initial deployments. Once VoLTE is everywhere, the whole idea of not having simultaneous voice and data will seem silly.

VoLTE also opens the door for cheaper LTE-only phones that can at last shed support for legacy voice, 2G, and 3G networks. This should reduce cost and complexity substantially, but LTE-only devices only make sense once LTE networks are 100% deployed.

For nomads – it is best to keep your options open and to make sure that you always retain the ability to fall back to older networks, which are often slow to get upgraded in some of the remote areas that we sometimes like to roam.

WiFi Calling

Similar to VoLTE, WiFi calling treats voice calls as data streams, but instead of sending them over the carrier's LTE network, the call is actually routed via local WiFi networks.

Not surprisingly, T-Mobile and Sprint have been the most eager to support WiFi calling, because it helps make up for their relative lack of cellular coverage. AT&T has also announced WiFi calling is coming "sometime in 2015" too.

WiFi calling opens up the door for using a cheaper cellular provider for voice service, and using a data-only connection on Verizon or AT&T to help ensure that you can still get coverage in more places, making sure that your voice calls and texts will still get through.

WiFi calling requires both support from your carrier, and compatibility built into your phone.

Understanding Cellular Frequencies

High-Tech Geek Alert:

This is a higher tech topic than most of the rest of the book, and we'll do our best to try to explain without getting too geeky. If this gets over your threshold of tech, just skip down to the last part of this chapter to make sure you understand what to look for when you shop for a cellular device.

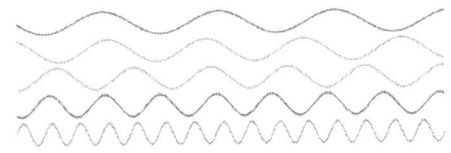

Every wireless broadcast travels along a radio wave with its own unique signature - its frequency.

The frequency of a radio wave is a measure of how many wave peaks there are per second - for example, a frequency of 700MHz means 700 million wave peaks per second are passing by.

The overall radio spectrum can be thought of as vast tracts of real estate along one very long road, with frequencies being the addresses of various properties, channels being marked off by lot lines, and bandwidth representing how much street frontage each channel owns.

A larger bandwidth channel means that a wider range of frequencies are used for a single transmission, allowing more data to travel faster.

The radio spectrum road ranges from the ultra-low frequencies where massive radios with miles-long antennas are used to communicate with submerged submarines, up to the extremely high frequencies used for exotic scanners and sci-fi-style energy weapons.

55

Lower frequency radio waves are better able to travel long distances and penetrate through walls and obstructions, and higher frequencies are better suited for higher speed transmissions.

The prime real estate in the middle of the radio spectrum is the UHF band (300MHz through 3GHz) – a sweet spot that allows for physically small radios that can still broadcast over relatively long distances.

This makes for a crowded chunk of spectrum, where digital television, WiFi networking, aeronautical navigation, car alarms, walkie-talkies, satellite radio, cellular voice and data service, and even microwave ovens all carve out their own valuable spot of real estate – squeezed in amongst dozens of other licensed and unlicensed users.

An Analogy: Let's say the cellular carrier are like a restaurants in business along this spectrum roadway: The more bandwidth a carrier owns translates to a larger building, thus more customers they can seat at a time. On this wireless information roadway, these channels of bandwidth are thus an extremely valuable resource.

Unfortunately, there is only so much of this prime real estate to go around.

Cellular carriers have spent billions of dollars buying up spectrum (and each other) so that they have the capacity they need, and in some major urban areas the networks are already operating at maximum capacity.

This is in part why carriers are so afraid of offering unlimited data plans – there just isn't enough physical capacity to offer an unlimited buffet to insatiable consumers. There is simply more demand for meals at their restaurants than there are tables to seat everyone.

A big part of the push towards faster networking technologies like LTE isn't just to offer customers faster speeds, it is to serve them faster so that a given chunk of spectrum can handle more simultaneous users. Just like a restaurant, the faster you can turn around the tables, the more customers a business can handle in a day.

As cellular data has grown more and more valuable and important, older areas of spectrum are being cleared out to open up more room.

- The 700MHz frequencies used by Verizon and AT&T's LTE networks used to be analog television channels 52–69, and this spectrum was only made available in 2009 as part of the transition to digital television.

- The 1700MHz/2100MHz AWS frequencies that T-Mobile relies so heavily upon were freed up by shutting down a wireless cable TV service that never really caught on.

56

- And right now the government is preparing to facilitate another big incentive auction in the 600MHz bands, allowing the current licensees of television channels 38–51 to voluntarily sell off their spectrum holdings so that they can vacate and make even more space for cellular data expansion.

Additional auctions are expanding the AWS and PCS (which are just the names of spectrum ranges, standing for Advanced Wireless Service and Personal Communications Service respectively) bands too.

The bulldozers are getting ready to move in, but it will still take a few years before the networks utilizing this newly cleared bandwidths are fully built and compatible devices are released.

Until then, the cellular restaurants remain crowded with customers willing to pay by the meal, so unlimited data buffets will be very hard to come by.

Cellular Frequency Bands

There used to be a time when there were only two neighboring chunks of spectrum dedicated to cellular usage, allowing for just two cellular carriers in any given region.

But things have gotten a lot more complicated as the move to 2G and digital opened up more potential channels, and Sprint pioneered this expansion by launching the first PCS network in 1995 on the 1900MHz PCS bands.

Now every carrier owns multiple blocks of addresses along the frequency roadway – but not all phones and data devices actually have radios capable of tuning in to every frequency and communications standard.

A smart shopper needing access to a carrier's entire network should make sure that his or her cellular devices support all the current and even the planned future frequencies used by their selected carrier.

Since everyone is standardizing on LTE, checking supported LTE bands is the simplest way to check devices for compatibility with various carrier networks.

Here are the current cellular frequencies in use in the United States:

Frequency	Common Name	Band	Who Is Using It?
600MHz	UHF TV Channels 38-51	tbd	Auction happening in 2016.
700MHz (Lower)	700MHz Auction Block A/B/C	LTE Band 12	US Cellular LTE, T-Mobile LTE
700MHz (Lower)	700MHz Auction Block B/C (Subset of band 12)	LTE Band 17	AT&T LTE
700MHz (Lower)	700MHz Auction Block D/E (Download Only)	LTE Band 29	AT&T LTE – Coming in 2015.
700MHz (Upper)	700MHz Auction Block C (Upper)	LTE Band 13	Verizon LTE
850MHz	Extended CLR (Cellular)	LTE Band 26	Sprint Spark
850MHz	CLR (Cellular – the original!)	LTE Band 5	Verizon 3G, AT&T 3G, AT&T 4G
1700MHz/ 2100MHz	AWS (Advanced Wireless Service)	LTE Band 4	Verizon XLTE, AT&T LTE, T-Mobile 4G, T-Mobile LTE
1700MHz/ 2100MHz	AWS-3 (Advanced Wireless Service - 2014 Auction)	tbd	Auction winners: AT&T, Verizon, T-Mobile, Dish Network
1900MHz	PCS (Personal Communications Service)	LTE Band 2	Verizon 3G, LTE, AT&T 3G, 4G, LTE, T-Mobile 2G, 4G, LTE
1900MHz	Extended PCS	LTE Band 25	Sprint 3G, Sprint LTE
2.3GHz	WCS (Wireless Communication Service)	LTE Band 30	AT&T – future LTE
2.5GHz	BRS (Broadband Radio Service)	LTE Band 41	Sprint Spark
5GHz	LAA (License Assisted Access)	tbd	Experimental: Verizon, T-Mobile

58

That is a lot of frequencies – and it is fiendishly complex to build a phone, booster, or any other wireless device that supports many of these at once. This is one of the reasons why many phones come in carrier-specific models and why roaming across carriers and internationally is often limited to the lowest common denominator frequency bands.

With all these incompatible frequency bands, things inevitably get messy – for example, the iPhone did not support T-Mobile's AWS 4G bands until the iPhone 5 came along. You could buy an iPhone 4S on T-Mobile, but you would only ever get 2G data speeds in most places - no matter what the coverage maps said.

Another example – US Cellular rolled out its LTE network in very nearly the same block of 700MHz spectrum as AT&T's. But very few devices support US Cellular's LTE Band 12, while almost all support AT&T's LTE Band 17. This means that in many areas, US Cellular customers are extremely limited in their selection of compatible phones if they want to get an LTE connection.

AT&T has promised the FCC to support Band 12 (a superset of Band 17) and to thus enable LTE roaming with a range of rural carriers, T-Mobile, and US Cellular. But it will take until the end of 2015 before this interoperability even begins to be fully implemented.

These are just a few examples of the growing pains going on out there.

It is almost miraculous that so many mobile devices work so well considering all the chaos behind the scenes.

Carrier Frequency Reference Guide

Verizon's Cellular Network

- **700MHz Upper Block (LTE Band 13)** – Verizon's primary 4G/LTE network.

- **850MHz Cellular (LTE Band 5)** – Currently supports voice calls, 2G 1xRTT, 3G EVDO. Eventual upgrade to LTE.

- **1700MHz/2100MHz AWS (LTE Band 4)** – XLTE (Extended LTE capacity in major metro areas)

- **1900MHz PCS (LTE Band 2)** – Currently supports voice calls, 2G 1xRTT, 3G EVDO. Migrating to LTE in major cities.

Verizon's Future Plans

Verizon has been rolling out what they are calling XLTE on the AWS frequency bands across their network to increase speed and capacity, and as of the beginning of 2015 most new Verizon devices are shipping with XLTE support.

In many areas Verizon's 700MHz LTE network is overloaded and slow while the XLTE bands are still uncrowded. If you do not want to get left behind, make sure that all your Verizon tech is XLTE compatible - and if you don't have XLTE support, an upgrade is very worthwhile.

Verizon has also begun to convert (aka refarm) their 1900MHz PCS network from 3G to LTE - and will continue to do so in key areas throughout 2015. In areas that have already been converted (like New York City) half of Verizon's 3G capacity has been turned off to make space for LTE - so if you are still using 3G you may start to notice speeds beginning to get increasingly worse.

Verizon's 850MHz Cellular network will continue as is for now, supporting older non-LTE devices.

Looking further ahead - Verizon invested heavily in the AWS-3 spectrum auction, and has begun experimenting with 5GHz LTE-LAA. No current Verizon devices yet support either of these LTE bands, but by the end of 2015 Verizon's future rollout plans may be revealed.

For maximum future compatibility – make sure that your Verizon devices support LTE Bands 2, 4, and 13 – and that they have support for Verizon's legacy CDMA network too. In the future LTE band 5 may begin to matter too.

AT&T's Cellular Network

- **700MHz Lower Block (LTE Band 17)** – AT&T's primary 4G/LTE network.

- **850MHz Cellular (LTE Band 5)** – Currently supports 2G GSM/GPRS/EDGE, 3G UMTS, and 4G HSPA+. LTE to be deployed here after 2G network is shut down.

- **1700MHz/2100MHz AWS (LTE Band 4)** – AT&T has begun to deploy LTE here in limited areas.

- **1900MHz PCS (LTE Band 2)** – Supports 2G GSM/GPRS/EDGE, 3G UMTS, 4G HSPA+, and 4G/LTE in select markets.

AT&T's Future Plans

AT&T has announced plans to fully shut down its 2G GSM network by 2017, meaning that older devices from before the 3G era will no longer work after this date. This freed-up network bandwidth will certainly go towards extending LTE capacity.

AT&T has already started to refarm its 1900MHz PCS network to support LTE, with several major urban areas converted.

AT&T has also been buying up 1700/2100MHz AWS spectrum, which it has begun using in limited areas, borrowing a feature from LTE-Advanced called carrier aggregation that lets AT&T combine channels from different chunks of spectrum to make a higher bandwidth virtual channel that can support even faster speeds.

Also taking advantage of carrier aggregation are AT&T's new download-only frequencies in the 700MHz LTE Band 29. These channels need to be combined via carrier aggregation with bidirectional upload/download bands to function, but it opens up a lot of extra capacity.

AT&T hasn't started using Band 29 spectrum yet, but has started to ship a few devices with support built in – meaning that these frequencies are likely to go into use sometime in 2015.

AT&T has also gotten FCC approval to expand LTE service into the 2300MHz WCS channels, but no current devices support these higher frequencies yet.

AT&T also invested heavily in 2014's AWS-3 auction, buying up more spectrum than any of the other bidders. It will take several years before the AWS-3 bands are ready for use however.

For maximum future compatibility – make sure that your AT&T devices support LTE Bands 2, 4, 5, and 17. Also keep an eye out for device announcements that support LTE Bands 29 and 30, and AT&T's carrier aggregation features for maximum future proofing.

61

Sprint's Cellular Network

- **800MHz Cellular (LTE Band 26)** – Supports voice and 2G 1xRTT, and 4G/LTE (Sprint Spark).

- **1900MHz PCS (LTE Band 25)** – Currently supports voice, 2G 1xRTT, 3G EVDO, and is Sprint's primary LTE band.

- **2.5GHz BRS (LTE Band 41)** – Supports 4G WiMAX and 4G/LTE (Sprint Spark).

Sprint's Future Plans

Sprint is pursuing deploying three LTE bands in key markets that, when combined, offer the best of all worlds – high-frequency speeds and low-frequency range. They are calling this tri-band LTE network Sprint Spark.

Though Spark is still in the early stages of being deployed, Sprint hopes to use the 800MHz spectrum for extended range coverage and building penetration, 1900MHz for mid-range, and multiple channels of 2.5GHz for extremely fast speeds in core areas.

Future Spark devices will offer support for LTE carrier aggregation, allowing Sprint to combine multiple channels together for increased speeds.

Meanwhile, Sprint's 4G WiMAX network has been at a technological dead-end for a while, and Sprint is currently phasing out WiMAX entirely – freeing up 2.5GHz spectrum to use for the Spark service. The WiMAX network is scheduled to be shut off completely by November 6th, 2015.

If you currently use a WiMAX-compatible device or one of the unlimited plans that includes only WiMAX as the 4G option, be prepared to be orphaned soon.

For maximum future compatibility – make sure that your Sprint devices are "Spark compatible" and support LTE Bands 25, 26, and 41. Also keep an eye out for devices that support carrier aggregation to allow for 2x and 3x 2.5GHz channels to be combined for even more speed, though none of the initial Spark-compatible devices support this.

T-Mobile's Cellular Network

- **700MHz Lower Block (LTE Band 12)** – T-Mobile's newest LTE band.

- **1700MHz/2100MHz AWS (LTE Band 4)** – Primary home of T-Mobile's HSPA+ 4G network and LTE service.

- **1900MHz PCS (LTE Band 2)** – Supports 2G GSM/GPRS/EDGE, 4G HSPA+ service in some areas. Being transitioned to LTE.

T-Mobile's Future Plans

T-Mobile has always suffered from a lack of coverage away from core metro areas, made substantially worse because T-Mobile has been relying on the relative high-frequency AWS bands, and has little low-frequency spectrum better suited for long-range reception.

To improve the situation, T-Mobile has recently bought up a lot of 700MHz LTE Band 12 spectrum, and in late 2014 began rolling out 700MHz service.

Without taking advantage of lower frequencies, T-Mobile will never be able to compete with Verizon and AT&T providing the long-distance rural coverage that is so important to RVers.

It is great to see T-Mobile beginning to fill the gap, but there is a catch - very few of even the newest mobile devices support LTE Band 12, leaving T-Mobile in the lurch until the device makers catch up.

Meanwhile - T-Mobile is also busy refarming its 2G 1900MHz PCS network to support HSPA+ and LTE too.

In the places it has coverage, T-Mobile has an extremely fast network. T-Mobile is labeling the places where they have deployed extra large bandwidth LTE channels as Wideband LTE markets, which are roughly comparable to what Verizon is calling XLTE. As of December 2014, T-Mobile has 121 Wideband LTE cities.

To help make up for areas where T-Mobile is lacking coverage, T-Mobile has a 7-year roaming agreement in place to allow for service on AT&T's 850MHz 3G network. This fills a lot of the gap for voice calls, but roaming data limits mean this is of extremely limited use for data.

Looking further ahead - T-Mobile invested in the AWS-3 spectrum auction, and is publicly gearing up to make a big play in 2016's 600MHz auction.

T-Mobile has also been leading the way in deploying voice over WiFi (VoWiFi) service, allowing calls and texting to be handled in areas where cellular is lacking. T-Mobile is also aggressively pushing forward with plans to begin testing 5GHz LTE-LAA service, witch expands LTE service onto the unlicensed spectrum normally used for 5GHz WiFi.

For maximum future compatibility – make sure that your T-Mobile devices support VoWiFi, and LTE Bands 2, 4, and 12 – and that they have support for 850MHz UMTS so that they can continue to roam onto AT&T. For future coverage LTE Band 12 support will be especially important, and at the moment extremely hard to find.

Frequency Asked Questions

Are some frequencies better than others?

Frequencies have trade-offs — lower frequencies travel farther and can more easily reach into the interiors of buildings. Higher frequencies tend to have more higher-bandwidth channels available, and thus offer faster speeds and increased capacity.

Carriers lacking in lower frequency spectrum have a much harder time offering coverage in remote rural areas since it takes a lot more cellular towers to cover the same amount of territory.

All the carriers are aspiring to build out multi-tiered balanced networks, with a mix of lower frequency spectrum for range and better building penetration, and higher frequency spectrum for speed and urban capacity.

But we are still in early days, and the race to bring out these enhanced networks is just beginning.

Why does the AWS band have two frequencies?

The AWS band is split into two chunks — with downstream data at 2100MHz and upstream data at 1700MHz.

All other cellular bands also separate uplink and downlink frequencies like this, but only in the AWS band are they so far apart as to merit having both frequencies written out explicitly.

Why do some carriers brag about having more LTE bandwidth?

LTE signals allows for 1.4MHz, 3MHz, 5MHz, 10MHz, 15MHz, and 20MHz bandwidth blocks to be devoted to data transmission — usually with a paired upstream and downstream connection described as 5x5 or 15x15 or 20x20.

Theoretical Peak LTE Data Rates

Bandwidth	Download Speed	Upload Speed
5MHz	37 Mbps	18 Mbps
10MHz	73 Mbps	36 Mbps
20MHz	150 Mbps	75 Mbps

The more bandwidth devoted to LTE, the faster the network can operate and the more users

64

that can be served simultaneously.

Many smaller carriers with legacy networks have only barely unleashed LTE, devoting just 1.4MHz or 3MHz of spectrum to LTE until they have transitioned enough users so that they can afford to take spectrum away from older standards.

Verizon, on the other hand, has been able to dedicate at least 10MHz to LTE in every market. AT&T only has enough free space to offer 5MHz universally, though in some places it has been able to expand to 10MHz and beyond.

As a rule of thumb – an LTE network can serve 200 active full-speed users for every 5MHz of bandwidth dedicated. If a network becomes overloaded, however, performance suffers for everyone.

Carriers bragging about 10MHz, 15MHz, or 20MHz channels are showing off how much headroom they have – and how much raw potential speed they can offer.

The LTE standard is limited to 20MHz bandwidth per channel, but the LTE-Advanced standard allows for up to 100MHz bandwidth – and for channels to be aggregated together by combining smaller channels from different noncontiguous chunks of spectrum.

How do I know a device I am buying fully supports my carrier?

Check the frequencies and standards listed on the specifications for the device to make sure that it supports all the current networks deployed by your carrier. Checking the LTE bands makes it easy, and be sure to check out our carrier reference guide in this chapter to be sure.

What happens if I buy a device that does not fully support my carrier's network?

Here's an example that shows why it is important to look closely at the specs - particularly if you are buying an older device:

T-Mobile's LTE network is primarily on LTE Band 4, their 3G/4G network is HSPA+ 1700/2100MHz, and their 2G network is 1900MHz GSM EDGE. Some T-Mobile areas have had the 1900MHz EDGE switched to 4G 1900MHz HSPA+, which is compatible with a wider range of devices. But many areas have not.

If you look carefully at the specs, you will notice that some popular devices (such as the non-Retina iPad Mini) are NOT compatible with T-Mobile's HSPA+ 1700/2100MHz 4G network, and neither is any Apple device older than the iPhone 5s manufactured before April 2013.

65

This means that in some T-Mobile 4G areas, non-Retina iPad Mini users will end up dropping back to super-slow 2G EDGE speeds.

The Retina iPad Mini, on the other hand, can handle every frequency that T-Mobile has in the US other than LTE Band 12, so Retina Mini users will often get 4G where Classic Mini users get 2G.

Only by knowing your frequencies will you be able to make a smart choice and avoid getting stuck unintentionally in the slow lane.

What if I want to roam internationally?

The GSM standard is by far the most deployed internationally – and quad-band World Phones are capable of roaming with at least 2G data speeds. Look for GSM 850MHz/900MHz/1800MHz/1900MHz support on your phone's or tablet's spec sheet. This used to be rare but is now very common.

LTE roaming is trickier – there are 44 different LTE bands in use around the world, and finding hardware that works on both your home network and where you want to visit may be tricky. And most US carriers have not worked out LTE international roaming agreements yet anyway. As of the beginning of 2015, only AT&T offers much in the way of international LTE roaming agreements.

If your plan is to find a local SIM, the latest flagship phones are the ones most likely to offer the most global LTE bands.

What about boosters frequencies?

Most older boosters are dual-band boosting – just the 850MHz cellular band and the 1900MHz PCS band. This typically helps with voice and basic data for all the major carriers but offers little help for the majority of LTE signals.

Some newer boosters are tri-band – coming in versions made specifically for Verizon, AT&T, or T-Mobile. In addition to the dual bands, the Verizon tri-band add support for the upper 700MHz band, the AT&T versions add support for the lower 700MHz band, and T-Mobile models add support for AWS.

Considering that Verizon and AT&T are also both rolling out service in the AWS band, getting a booster that lacks this support might prove to be short-sighted – though presumably anywhere Verizon and AT&T have AWS networks, they will also still have LTE in the 700MHz bands.

There are, however, some five-band boosters that offer support for all of the above. If you have devices from multiple carriers, a five-band booster is especially wise. Sprint is the odd duck out – no current or announced cellular boosters fully support Sprint's tri-band Spark scheme.

What on Earth is refarming?

The curious word "refarming" refers to the practice of shuffling around existing network frequency allocations to make room for new services and more efficient modern technologies.

In the process, older technologies end up with slower speeds and less coverage allocated to them, or service may be eliminated entirely.

If you happen to still have an old device that is going to be made obsolete and unusable by network refarming, carriers have been known to offer free hardware upgrades to keep from losing a customer.

WiFi & Hotspots

Often the fastest, cheapest, and easiest way to get online is to use public WiFi networks, and these are growing increasingly easy to find.

Many libraries, coffee shops, RV parks, breweries (yay!), motels, municipal parks, and even fast food restaurants now offer free WiFi. There are also plenty of paid WiFi networks to be found too.

Though WiFi has the potential to be blazingly fast, some shared networks can be horribly overloaded. A public WiFi hotspot is highly dependent upon their upstream source of internet (cable, DSL, satellite, etc.) and on how many people are sharing that connection.

WiFi Hotspot

Quick Glance

Pros

Widely available

Easily accessible

Frequently free

Cons

Variable quality

Frequently unreliable

Security concerns

In some cases, the upstream connection may actually be little better than old dial-up modems. In some remote places, the upstream connection may actually BE a dial-up modem!

Unfortunately – in many situations, even though you may be able to get online via WiFi – it may not be worth bothering with.

The other major limitation of WiFi is range. Sometimes we enjoy working in a cafe or brewery, but usually we prefer to be at our home office or computing outside under the shade of a tree. Most WiFi hotspots fall off to unusably slow connections just a hundred feet away from the base station, and in some RV parks only the nearest spots to the front office can reliably connect via WiFi.

But with an external WiFi antenna and/or WiFi repeater system, you can often manage to connect to a base station substantially farther away than your unaided laptop or tablet alone ever could.

WiFi Connectivity Options

Generally, there is a rough hierarchy of WiFi transmission strength and receive capability – from weakest to most capable:

1) **WiFi Gadgets** (including phones, tablets, etc)

2) **Laptops**

3) **Indoor WiFi-as-WAN Routers** (WiFiRanger Go, Pepwave Surf SOHO, etc.)

4) **High-Power USB WiFi Network Adapters** (Alfa, etc.)

5) **Outdoor CPE / Router** (WiFiRanger Sky, WiFiRanger Elite, Wave WiFi Rogue Wave, The Wirie AP+, etc.)

6) **Outdoor CPE / Directional Antenna** (Ubiquiti NanoStation, parabolic dish, etc.)

Most everyone knows what a laptop is, but as you go down the list from there things start to become more and more exotic sounding.

But they are actually rather simple once you understand a few key WiFi-related terms:

- **Router** - A router connects to an upstream "wide area network" (WAN), and provides service to multiple downstream devices on the local area network (LAN) by routing where the network traffic goes. A home router is usually plugged into a cable or DSL modem for the upstream WAN connection, and a mobile

70

hotspot is just a specialized router that uses a cellular network upstream.

- **WiFi-as-WAN** - Some WiFi routers support using another WiFi network as the upstream WAN - using a single WiFi radio to connect to the WAN and create a WiFi LAN simultaneously. This is a rare feature in typical off-the-shelf home routers, but is especially valuable for RVers.

  Since a WiFi-as-WAN router will usually have a stronger radio and better antennas than any of your local devices, a WiFi-as-WAN router can act as a relay, letting devices on your LAN connect to a WAN further away than they ever could alone.

- **USB Network Adapter** - You can't improve the WiFi built into your laptop, but it is possible to add a second more powerful external WiFi radio via USB. A high powered USB network adapter with an antenna placed in a window can outperform an indoor WiFi-as-WAN router, but it can only help in getting a single laptop online. It can't help with multiple devices connecting at once, or with connecting any devices without a USB port.

- **CPE** - CPE stands for "customer premises equipment" and is simply the common term used to describer commercial-grade outdoor-rated WiFi access point used by wireless service providers. A CPE combines roof-mounted height, better antennas, and a more powerful radio to deliver the maximum possible WiFi range.

  The WiFiRanger roof mounted Elite, Sky, and MobileTi, The Wirie AP+, and the Wave WiFi Rogue Wave products are all actually commercial-grade CPE's under the hood with custom user friendly simplified interfaces tacked on top.

  Some CPE's can be configured as WiFi-as-WAN routers, creating a local WiFi LAN. Others CPE's just act as an access point, and need to be paired with an indoor router to make a local WiFi network for your devices to connect to.

For more on how to use and improve your WiFi connections, see the chapter on "Enhancing the Signal."

Only once you've digested your options and your needs will you be able to decide just how much investing in long-range WiFi is worth it to you.

Security on Public WiFi Networks

The internet can be a scary place – it is important to be careful out there!

On a public WiFi network, you should assume that anyone else connected to the same network as you has the potential to eavesdrop on your communications.

You may not think that any of the other campers nearby appear to be hackers, but...have you considered the possibility that some of them may have been hacked in the past?

Many computers that have been infected by malware have stealthy spy software running on them, looking for other computers to infect or potentially valuable data to log. The friendly grandma and grandpa with the cute doggie camped at the next spot over may not care that you're online paying your bills, but their obsolete Windows XP laptop might be working for the Russian Mafia – logging everything that happens nearby.

The only sure way to protect all of your communications is to use a VPN (Virtual Private Network) service that encrypts everything between your computer and the VPN provider's central office, bypassing any potential eavesdroppers nearby.

But – is it worth the trouble, decreased speed, and cost to encrypt reading the day's news and checking the weather? Much of what most people do online isn't worth protecting.

A lot of websites are secured using SSL (Secure Socket Layer), which acts like a VPN between your browser and that particular site, effectively blocking anyone from listening in without the computational resources of the NSA to devote to decryption.

In other words, if you load a website using a URL starting with "https://" or your browser displays a padlock symbol – you are generally secure logging into that site, even on public networks.

Almost all financial sites use https, and – after some very embarrassing hacks – even social networks like Facebook and Twitter have begun to allow "https://" connections.

Always use them if a site gives you the option!

But very few blogs or online forums give you the option of using https:// to secure your connection, as it significantly increases the cost to run the site to do so.

Your biggest actual risk when using a public WiFi network is not when you log into a secure site like your bank but when you log into insecure websites and online forums using the same username and password combinations that you have also used elsewhere. In this case, it is all too easy for other machines on the same public WiFi network to "sniff" your password and login credentials.

A hacker isn't likely to want to impersonate you on an RVing forum, but he or she will take that password and use that to try and get into other critical secure sites. If you have used the same password and username anywhere else online, your identity is at risk!

> The only way to protect yourself is to pay attention to the number one rule of Internet security – **NEVER ever ever use the same password twice. DON'T DO IT!**

It might seem ridiculous to try and memorize different passwords for every site you visit, but you don't have to. Use a password management program to do it for you. There are several great options out there – we personally use 1Password (agilebits.com/onepassword), which syncs all of our password files between our laptops and mobile devices.

Or at the very least, keep your key passwords written down in a private paper journal – with a photocopy saved somewhere safe in case of disaster.

Hackers don't like physically breaking in to get things – they can attempt billions of logins an hour (really!) on a basic home PC with some fancy cracking tools, but physically breaking in someplace is messy and time consuming with high risks.

So keep this always in mind – the odds are a lot higher that the RV next door has a laptop listening in for passwords sent to insecure sites to report to the Russian Mafia than it is that a shady hacker will physically break into your RV and discover the sheet of passwords printed out and hidden deep in your glovebox.

> **Security update:** If you are still running Windows XP, stop reading this book right now and upgrade that machine ASAP.
>
> XP has reached the point where it is no longer being supported with security updates from Microsoft, and the core security in XP is long obsolete. At the same time, the Russian Mafia is still very actively supporting Windows XP, and hacker tools for exploiting XP systems are thriving.
>
> By taking every remaining XP machine off the network, you are doing not only yourself – but everyone – a favor.

73

VPN Services – Protection, Privacy, and More

Virtual private networks (aka VPNs) encrypt all the data between your devices and a central VPN server – so that no one on your local network can tell where you are surfing.

In the past, VPNs used to be strictly for advanced users who knew how to manage technical configurations and who had access to their own central server to act as a host.

Now, however, there is a wide range of VPN providers offering incredibly simple-to-use services with clients for Mac, PC, iOS, and Android. No advanced configuration required – just click "On" and your connection is protected.

Some highly regarded easy-to-use VPN services to consider are:

- Spotflux (www.spotflux.com)

- Cloak (www.getcloak.com)

- TunnelBear (www.tunnelbear.com)

And some geekier, more advanced options:

- Private Internet Access (www.privateinternetaccess.com)

- AirVPN (www.airvpn.org)

Some VPN services do a lot more than just encrypt your traffic. One of the coolest features many offer is the ability to change the central VPN server you are communicating through – in essence, changing where you appear to be on the internet.

This location-changing feature allows you to appear to be connected from a different country – allowing you to stream BBC shows as if you were in the UK, for example. Or if you are traveling outside the US, with the help of a VPN, you can still get access to Netflix as if you were still at home.

Some VPN services also offer additional connection optimizations, such as ad blocking or automatic data compression. Shop around and try a few free trials to decide what most appeals to you.

VPN services usually offer limited free service plans, or cost $30–$120 a year to secure all of your devices.

74

Watch What You Share

One more critical and often overlooked security tip for connecting to public WiFi: Be aware of what you're sharing publicly on your computer to other users on the local network!

Once on a RV park's public WiFi network we were able to see and access photos of our neighbor's dog that she was sharing publicly in iPhoto. Obviously, that's probably not a big deal – her dog was cute and brought us a smile. But what if she had been inadvertently sharing a photo library more...umm...personal?

We also frequently discover people's shared music libraries, movies, and documents.

Know how to turn public sharing on and off in your operating system and key applications, or you might risk sharing something much more embarrassing and personal than your Taylor Swift music library!

If you are not sure what you are sharing, ask a trusted friend on the same network to look for your computer and try to connect. If they can't even see you, you are golden.

And have you ever considered what information your phone or hotspot is giving away about you? Even a password-protected network might reveal way more than you'd imagine – just by what it is named.

We recently had some campground neighbors who's personal hotspot name is "John Doe's iPhone" (name changed to protect the clueless). His network was password protected and secure, but his name is rather unique. Two seconds in Google reveals that he is 55–59-year-old male, and we even got his home address and phone number.

With ill intentions, who know's what might be possible – particularly knowing his home address and that he has taken his family away for vacation.

The lesson: If you want to remain anonymous to your neighbors, take care what you name your network, and don't rely on the default auto-generated name your phone may use when creating a hotspot.

And if you want to have fun with your fellow bandwidth-hungry campers, come up with clever network names like "FBI Surveillance Van," "Have a nice day :)," or even broadcast your travel blog with something like "technomadia.com" as a way to make new friends!

75

Realities of Campground WiFi

Although you would think that a campground that advertises "Free WiFi!" as prominently as it does 50A power hook-ups would actually have worthwhile WiFi, we have sadly discovered that this is often not the case. And once you understand all that is involved in providing such a service as free WiFi, you may realize how unrealistic it is to expect that.

Maintaining a public WiFi system that can serve hundreds of bandwidth-hungry travelers, especially if spread out over several acres (such as at an RV park), is very expensive to set up and maintain. Few RV park managers have the expertise to upkeep such a network, especially without having an IT manager on call.

To do it right requires a substantial investment in routers, repeaters, and equipment – not to mention, needing a pretty hefty backbone of internet bandwidth to tap into.

Generally, if the WiFi is managed decently enough, it is common for RV park WiFi to be good enough for checking email and doing some basic surfing – generally all that most RV campground patrons are assumed to really need.

But if all you are going to do is check email and some light surfing – it is hardly worth trying WiFi when a cellular plan could do it just as well.

On rare occasions, we've been at campgrounds with great WiFi – fast and without data caps! It has occasionally been so good we could suspend our cellular account for the month.

But it is much more common for the signal or configuration at campground WiFi installations to be so iffy that it's pretty much unusable for anything.

All too often, campgrounds suffer seriously overloaded connections – particularly in the evening when a lot of people try to get online at once. A connection that might be decent during the day, while everyone is off exploring, might feel worse than dial-up during prime time. It can take just one or two people trying to stream a movie, video chatting with the grandkids, or downloading a huge file to bring the entire network to a grinding halt.

Some campgrounds have outsourced the chore of providing WiFi to a provider, like Tengo Internet (www.tengointernet.com), who manages the bandwidth and network for them. Sometimes they even charge extra for it, cap how much data you can use, or limit how many devices you can connect at once.

The theory is that these limits help more fairly spread out the bandwidth.

You would think that paid and professionally managed WiFi connections would end up being faster and more reliable than open and free, but in our experience this has not proven to be the case. All too often, the paid networks have proven to be a waste of effort, not even worth bothering to jump through the hoops to get online.

In the end, we look at campground-provided WiFi to be a bonus if it's usable – but we've learned not to rely on it for anything critical and bandwidth intensive.

If campground-provided WiFi is important to you, read online reviews at places like campendium.com, www.rvparkreview.com or www.rvparking.com to help in selecting where you head – a lot of folks comment on the reliability of the WiFi in their park reviews. And if you find the WiFi not cooperating, try asking the front desk to reboot the router – that's a simple thing that can sometimes make a huge improvement.

Free Public Hotspots

A lot of businesses and public resources have begun to provide free WiFi. For example – buy a cup of coffee, and you can spend an afternoon using a cafe's hotspot for your laptop.

Heck, even many McDonald's offer free WiFi to go with your super-sized fries. "Would you like bandwidth with that?"

If you have the flexibility and/or desire to take your laptop with you, these can be great supplements to your internet arsenal. Some folks actually rely on this method as their primary internet. And if you can get your RV within range of these hotspots, you might even be able to use them from your home on wheels.

Places known for their free hotspots include libraries, McDonald's, coffee shops, Panera Bread, Lowe's (yes, the hardware store...many even offer free overnight parking too!), some rest stops, motels (higher end hotels tend to have paid internet), breweries, restaurants, and so many more.

There are several apps and websites out there that you can use to track public hotspots down – but we usually find it pretty easy to stumble into these places when we need them.

Sometimes you might have to ask for the password, but many places are happy to share their bandwidth with their customers.

Some days we find a restaurant with free WiFi to eat lunch at, and then spend the afternoon while the tables aren't in high demand working away. Some places are even happy to seat you near a power outlet!

Some other places limit how long they'll let you stay on their connection, as they do need their tables for customers just arriving. If asked to move on, be courteous and comply.

These businesses provide these services as a courtesy to their customers – please return the favor by being their customer, ordering food, and tipping well.

And always remember, not all WiFi is created equal and your speeds may vary from location to location. Before buying a meal or coffee, we run a speed test from our devices to make sure the bandwidth offered is usable enough for our needs.

Paid WiFi Networks

There are some widely deployed WiFi networks out there available to paid customers.

For example - Florida cable company Brighthouse is installing a network of WiFi hotspots across Central Florida for their customers – and this is part of a nationwide Cable WiFi (www.cablewifi.com) initiative involving other regional cable companies (including Time Warner Cable, Cox, Optimum, and Comcast's Xfinity), all of which allow customers to roam freely between connected Cable WiFi hotspots.

With over 300,000 WiFi hotspots and growing, if you are in an area served by one of these companies you might be surprised to find that you can get access in some very unexpected places – if you are an authorized user.

Check if your stationary family members and friends are customers of one of the participating cable companies – you might be able to get their permission to use their login to access the network.

Comcast's Xfinity WiFi (wifi.comcast.com) is taking things even further – and now has millions of "xfinitywifi" hotspots. Comcast has accomplished this by turning business and home customers' cable modems into public WiFi hotspots – in one fell swoop offering fast WiFi over entire neighborhoods.

Non-Xfinity customers can get two 60-minute complementary WiFi sessions a month, or you can buy hourly passes for $2.95, daily for $7.95, and weekly for $19.95.

If you see "xfinitywifi" or "CableWiFi" as an available hotspot, you can try this out as an option. If you are in a Comcast area, the Xfinity iOS and Android app will help you find areas that are covered.

There's also services like Boingo (www.boingo.com), which has over one million hotspots around the world – including many Tengo location found within campgrounds. For a low monthly fee of just $9.95, you can get unlimited access within the Americas for up to two devices. This can oftentimes be cheaper than a Tengo campground pass for the same service.

Boingo also offers a global plan for international travelers.

Borrowing Bandwidth From Friends

Our favorite way to access WiFi is by borrowing a cup of bandwidth from friends and family as we travel.

We find most folks with fast home connections are more than happy to share their unlimited high-speed bandwidth when we need to do things like OS updates, download shows from iTunes, get the latest development tools for our software projects, or do a massive backup to DropBox. (All things we avoid doing over cellular!)

We typically keep small meaningful gifts on board to thank our gracious hosts.

Enhancing the Signal

There is nothing on Earth more frustrating than the agony of having just one bar of signal, whether that signal happens to be WiFi or cellular.

That solitary bar is a cruel tease — usually not enough signal to actually reliably use, but it's there, taunting you.

If you maybe just hit reload one more time…or shift your position a bit…

Right at the moment you are about to give up, it works! For a few minutes, at least. Just long enough to keep you on the hook trying.

At least when there is no signal at all, you can concentrate on doing other offline things. But having a hint of signal…that is the path to madness and not accomplishing anything.

There are, however, things you can do to improve a bad situation – taking a weak signal and making the most with it. With external antennas, cellular boosters, and WiFi repeaters, the results can sometimes be near miraculous.

But other times, no amount of boosting can help – that cruel solitary bar might remain undefeated, taunting you still.

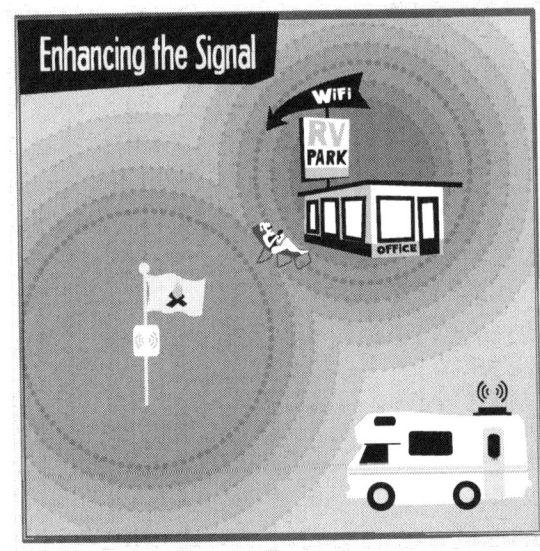

81

Things That Impact Wireless Signals

Wireless signals can be quite variable and are influenced by so many things. Sometimes just moving from one side of your rig to the other can make a dramatic difference.

In general, there are a few universal things that help with reception, and these guidelines are relevant for both cellular and WiFi.

Things that help:

- **Line of Sight:** Nothing improves a wireless signal more than having nothing between the sender and receiver. If you can visually see the cell tower or the WiFi hotspot, there is a good chance your wireless device can too.

- **Altitude:** The best way to get line of sight, or at least fewer obstructions, is to get up over the clutter. An antenna at the top of a 20' mast has nothing but clear air around it. An antenna mounted directly on roof might be blocked when a giant fifth-wheel moves in next door, or even by your own air conditioner. And an antenna stuffed into the back of a crowded lower cabinet is starting off with a substantial handicap.

- **Directional Focus:** A transmitter of a given loudness (power) can be heard a lot farther away if the energy is focused towards the receiver – just as speaking into a megaphone does wonders to project your voice. Of course, the downside of having a directional antenna like this is that it adds a manual aiming step. Since the signal off to the sides outside of the sweet spot is substantially diminished, directional antennas are often more trouble than they are worth if you're just passing through. Only in a relatively fixed location is the extra effort worthwhile.

- **Power:** Louder is better – but only up to a point. If you push too much power out a radio antenna, it can begin to overwhelm your target. And worse, it can drown out others on the same or nearby frequencies.

- **A Sensitive Listener:** Just like in relationships, often better than raw power is a sensitive ear. If the receiver is listening carefully, even a weak signal might be heard. This sensitivity is often a prime difference between expensive commercial-grade wireless gear and cheap consumer-grade stuff.

- **An Uncrowded Channel:** Wireless networking devices are designed to allow multiple users to share a single channel. On a cell tower or a crowded campground WiFi network, there could be hundreds of devices all trying to talk at once. The more people talking at once, the more congested and degraded the network becomes for everyone.

82

- **Upstream Capacity:** A river can't flow through a straw – and often the real problem with both campground WiFi and cell towers is there's only a straw's worth of pipe bringing in the water. Some campgrounds invest in great WiFi equipment, and you might have the strongest signal ever – but upstream they have skimped and have little more than a single basic DSL line serving the entire campground. The same applies to cell towers, particularly in more remote areas. The tower may speak fast LTE, but if the upstream network (called the backhaul) is not sufficient for the population being serviced, the actual speed you see may be extremely limited.

- **Antenna Diversity:** Having multiple antennas working together is called antenna diversity, and this can work wonders with compatible equipment – particularly in indoor areas where there may be a lot of signal reflections bouncing around. A secondary antenna can compensate for a primary antenna being in signal shadow.

There are some universal things that degrade signals – both cellular and WiFi. Things that hurt:

- **Metal Obstructions:** Barriers between you and the WiFi base station or cell tower are bad. But metal barriers are very, very bad. Radio waves in general have difficulty passing through metal, making getting a signal for those living inside metal buses or shiny Airstreams an extreme challenge.

- **Interference:** The more noise on the line, the harder it is for a radio receiver to hear. Noise can originate with other intentional signals or it could be background noise from microwave ovens, hair dryers, or even the sun.

- **Overcrowding:** Overcrowding leads to a vicious spiral – with too many radios trying to broadcast on one channel, it becomes hard to pick out individual conversations. Transmitters try to compensate by broadcasting louder and repeating themselves to get a message through. That leads to even more overcrowding and noise – until eventually the wireless network saturates and barely any data is making it through.

- **Power:** Some WiFi devices and routers let you manually set your radio power. It seems counterintuitive, but you can actually sometimes improve both speed and even range by dropping down to a lower power setting. At full power, you may be too loud, and you could be generating interference and overcrowding your neighbors. Speaking in a whisper is sometimes more effective than roaring at the top of your lungs.

A lot of these things may be out of your control – but it helps to understand the things that may be impacting your signal in any given situation.

Overview of WiFi & Cellular Options

If you've got a little bit of signal and want more, there are ways to go about trying to get it that don't involve climbing trees or driving into town.

The techniques are different, however, for WiFi and cellular. Here is an overview:

WiFi Signal Enhancement

- **USB WiFi Access Points / Network Adapters:** The WiFi radios in most laptops and mobile devices weren't designed with maximum range in mind. One of the most affordable ways to get a bit more range is to use a USB WiFi access point that has a more powerful WiFi radio in it. With a USB extension cable, you can stick the access point into a window facing towards the target hotspot – and some WiFi adapters even have antenna ports that can be used with outdoor antennas too.

  The Alfa 1000mW 802.11b/g WiFi network adapter (available from TechnoRV.com) has built up a reputation for delivering great performance at range, though it has not been updated with support for the most recent operating systems and WiFi standards.

  The downside of an adapter like this is that it only gets one computer online, and juggling the wires can get awkward.

- **Indoor WiFi Repeaters:** A WiFi repeater goes a step further and combines an access point with a router. The access point picks up on a remote WiFi signal, and then the router creates a new local hotspot that all your devices can connect to. For much more on this option, see the next section of this chapter on WiFi as WAN.

- **Rooftop WiFi:** Particularly with campground WiFi, a good signal is often very hard to find at ground level. Some RV WiFi gear – like the WiFiRanger Sky & Elite, the Wave WiFi Rogue Wave, and The Wirie AP+ – are actually commercial-grade CPE radios designed to be installed permanently outdoors directly onto the roof of an RV, maximizing range. The upfront cost may be steeper, but this is where the most substantial gains are to be had.

  Some of the same benefits can be achieved by using an indoor router as a repeater, connected to high-quality external WiFi antennas on the roof. But for every foot of wire running between the outside antenna and the indoor router, much of the potential gain is lost.

Cellular Signal Enhancement

- **External Antennas:** Very few phones and tablets have antenna ports anymore, but some mobile hotspots and USB modems do — and all commercial-grade equipment does. If you have an antenna port, you can attach a more capable antenna for better reception. This is especially helpful if you want to keep your mobile hotspot set up in a fixed central location but allow for an antenna in a window or on the roof.

  Some cellular devices support dual external antennas. Installing a matched pair of antennas preserves antenna diversity and allows for LTE's multiple-input, multiple-output (MIMO) capabilities. Diversity provides for increased performance in noisy signal areas with a lot of signal reflections, and MIMO allows for multiple data streams to be combined for greatly enhanced speed. If your equipment supports it, the performance improvement from a dual diversity antenna installation can be dramatic.

- **Cellular Boosters:** A booster picks up a signal with an external antenna, amplifies it, and rebroadcasts it via an internal antenna. The maximum legal gain for a mobile cellular booster is 50dB, and for a home booster the maximum gain is 70dB. Boosters are frequency specific — you need a booster designed to work with the frequencies that your carrier uses. (See the "Understanding Cellular Frequencies" chapter for more.)

- **Smart Boosters:** A new type of smart booster actually functions more like a WiFi repeater, only for cellular. Smart boosters are carrier specific: Rather than just indiscriminately amplifying a raw signal, they know enough about the carrier's network to capture a signal and recreate it indoors. Because the booster is recreating the signal and not just amplifying it, it is able to ignore the noise — allowing for as much as a 100dB gain. Smart boosters are at the moment only appropriate (and licensed for use in) fixed homes, but keep an eye out for possible future RV-friendly implementations.

  Cel-Fi (www.cel-fi.com) is the most prominent company pushing smart booster technology, and they have told us a mobile version is on their longer term radar.

WiFi as WAN & WiFi Repeating

When a WiFi signal is just too far away to reach from your RV — this is where WiFi repeaters can help out.

The capability of a WiFi router to use a different WiFi network as its upstream connection is known as WiFi as WAN, or WiFi Repeating.

Most mainstream consumer WiFi routers do not support WiFi as WAN – but WiFiRanger, Pepwave, and Cradlepoint are all examples of router brands that support this as a core feature.

Having a roof-mounted WiFi access point paired with an indoor WiFi router is especially handy if your RV is a metal signal-blocking tube, like our bus is. Getting a transmitter up on the roof means that we don't have to fidget with our gadgets trying to figure out which window has the best signal today.

We currently have both a WiFiRanger (www.wifiranger.com) Sky and WiFiRanger MobileTi on our roof, and have it paired with a WiFiRanger Go2 inside our bus. The Sky picks up the remote WiFi signals, and the Go then creates a local secure WiFi and wired ethernet network for all of our devices.

When we have WiFi access, it works great. When we don't, the WiFiRanger is also cellular aware so we can – with just one click (or via automatic failover) – switch over to our cellular hotspot, keeping all our devices connected with minimal interruption.

Another excellent roof-mounted WiFi system we have heard good things about is The Wirie AP+ (www.thewirie.com), created by full-time sailing nomads.

Though not rated for outdoor use, Pepwave and Cradlepoint routers can be installed indoors and utilize outdoor antennas for a very capable solution.

Roof Mounted & Directional CPEs

One little-known yet totally awesome feature of the WiFiRanger routers is that they support a feature called WFRBoost that lets them remotely control and manage many Ubiquiti CPE devices.

In our arsenal, we have a Ubiquiti NanoStation 2, a small affordable (less than $100) CPE with a built-in directional antenna designed for pole mounting.

When plugged into a WiFiRanger, it shows up as just another signal source in the main WiFiRanger control panel, seeing networks vastly farther away than the WiFiRanger Sky or MobileTi ever could.

The downside is the setup time -- the NanoStation's directional antenna does wonders pulling in a distant signal, but you need to spend time slowly rotating and checking signal strength until you find an optimal setup.

We have managed to connect to open WiFi networks over a mile away by carefully aiming the NanoStation – and in the right conditions, ranges even farther are possible if the target base station is also up to the long-range task.

We don't consider the NanoStation a replacement for our WiFiRanger MobileTi, but rather a complement to it.

The MobileTi is better for medium-range and passive situations, and doesn't require setting up a mast or aiming an antenna to use. But when we are stopped someplace for a bit, we've found it occasionally worth the effort to fine-tune the NanoStation.

An external CPE like a NanoStation can also be used with other routers, but there is substantially more manual configuration involved.

The Wave WiFi Rogue Wave (www.wavewifi.com) however is an excellent roof-mounted CPE with a very friendly end-user interface worth checking out.

Cellular Boosting

One of the problems with cellular internet is that the signal can be quite variable depending on many factors – your device, the tower location, weather, how many people are also using the tower, local terrain, nearby buildings, and even your own RV's construction.

You can buy various antennas and boosters that can improve the situation. Sometimes these are lifesavers and can make a finicky signal usable enough to get your work done.

However, these are not miracle devices – they can't make signal out of nothing. There has to be some signal nearby for them to work with, and even then it may take some tweaking. A cellular booster works by taking a weak signal, amplifying it, and rebroadcasting it to a room, RV, vehicle, or building.

When you are on the edge of a coverage area – or even dipped down into a canyon where cellular signal may not reach as strongly – this can work wondrously.

If you are someplace where the signal isn't already weak or where there are others nearby trying to get connected, a badly designed or improperly tuned booster is akin to a drunken loudmouth ruining the ambiance by shouting across the table in an intimate restaurant.

The potential for interference is real. And to the carriers who have spent billions of dollars buying up the rights to cellular spectrum, a consumer-grade booster stepping on their toes is not at all welcomed.

87

But at other times, carriers themselves are eager to recommend boosters if it helps them sell service to customers in fringe areas.

In early 2013, things at last came to a head, with the FCC working with cellular carriers and booster manufactures to jointly come up with new standards to prevent interference – yet to still allow boosters to be sold without needing special permission granted to each consumer.

The new rules went into effect in May 2014, and new approved boosters are finally starting to make it to market.

Some of the key manufacturers of these devices include:

- weBoost (formerly known as Wilson Electronics) (www.weboost.com)

- SureCall (www.surecall.com)

- Top Signal (www.topsignalproducts.com)

- Maximum Signal (www.maximumsignal.net)

These devices can range from simple systems that just boost one mobile device at a time to complex systems that can provide RV-wide boosts for multiple devices simultaneously, even on multiple different carrier networks.

There are, as of yet, limited options for systems that can boost LTE devices, but more should be coming to market soon. And keep in mind many boosters may support your carrier but not all the frequencies your carrier uses.

For advice on these systems, we recommend the folks at PowerfulSignal.com and 3GStore.com, both of whom have been extraordinarily helpful and have a specific understanding of RVing needs. The devices are also available for purchase at retailers like Amazon.

Keep Current Alert:

We keep this Resource Center page updated as the cellular booster options change - start your shopping here for specific models:

Comparison: Mobile 4G Cellular Boosters

(www.rvmobileinternet.com/resources/mobile-cellular-boosters/)

(Our Mobile Internet Aficionados members also get access to our in-depth testing reports.)

Cellular Signal: Bars & Dots vs. dBm

Everyone knows that more bars is a good thing – but very few people realize that different phones and operating systems calculate how many signal bars to display very differently.

This means that comparing bars, unless you are on the same phone and same carrier, is actually a very poor way to compare coverage and signal quality between different devices.

The bars your phone is displaying sometimes do not even directly correspond to the actual signal strength: In addition to raw signal strength, the phone may be measuring network congestion and other variables to calculate how many bars to display.

In general, iPhone puts more weight on network congestion when calculating how many bars (or dots, as of iOS 7) to display, and Android focuses more on raw signal strength.

This can make it harder to measure the impact a booster or antenna is having, because improving the signal strength might not register as more bars if the phone is focusing on network congestion.

To objectively compare signal strength, you need to look for the dBm reading, which is based upon the actual power in milliwatts being received by an antenna.

This is a logarithmic scale – a 0dBm reading represent a single milliwatt received, and every change by 10 up or down represents a 10x change in received signal power.

The power levels picked up by a cellular antenna are fractions of a milliwatts, so the dBm readout will be a negative number.

> –50dBm would be considered an awesome signal.
> –60dBm is 10x weaker, but still great.
> –70dBm is 100x weaker.
> –80dBm is 1,000x weaker.
> –90dBm is 10,000x weaker.
> –100dBm is 100,000x weaker – and is when you are likely to start seeing a serious impact.
> –110dBm is a million times weaker than –50dBm and is usually barely usable.
> And by the time you see –120dBm, the phone has probably already given up and switched to "No Service."

Modern radios can work wonders with weak signals. In the past, −95dBm would have been considered weak, but a modern LTE radio can work with signals of −100dBm all the way down to −110dBm and occasionally beyond.

Personally, we are blown away that a tiny, relatively affordable gadget that fits in your pocket can pick up signals so incredibly weak and do such amazing things with them.

The science involved here borders on magic.

See Real Signal Strength on an iPhone or Android

A hidden feature found in every iPhone running iOS 4.1 or newer is the ability to enable a special Field Test Mode that displays a real signal strength. This mode is enabled by going to the dial pad, and dialing *3001#12345#*.

On many Android phones, this reading can be found by going to Settings -> About Phone -> Status.
If it's not there, you may need to dig around.

Cellular Booster Tips & Tricks

Here are some tips and tricks for getting the most out of a cellular booster:

- **Boosters Boost Battery Too:** Nothing drains a cellular device's battery faster than being in a fringe signal area. To stay connected, your gadgets need to operate their transmitter in full-power mode, and they are constantly searching for a better signal to lock onto. If a cellular booster is in the mix, your gadgets can transmit in low-power mode – vastly improving battery life. If you know you are going to be in a fringe area, put your phones into airplane mode. The battery will thank you! (Ever wonder why, when driving through more remote areas, your phone's battery seems to drain quicker than normal? This is why!)

- **Know the Frequencies:** Even a booster sold as compatible with your carrier may not support all the frequencies your carrier uses. Know what your booster supports and what your carrier uses – and set your expectations accordingly. (See the "Understanding Cellular Frequencies" chapter.)

- **Toggle Airplane Mode after Enabling a Booster:** Your mobile device occasionally may not notice that a new stronger signal has suddenly become available once you turn a booster on. To make sure that it does a full scan and finds it, after you turn on your booster it can help to reboot your device or toggle it into and then out of airplane mode.

- **It's Not Just About Bars:** Especially on iOS devices, the signal bars displayed reflect network congestion as well as raw signal strength, and as such, enabling a cellular booster may not show up as more bars even if the signal is now stronger. Instead of relying on bars, run a speed test to see what the actual difference in connection speed is.

- **Uploads Are Most Impacted:** Sometimes a booster may barely improve your download speed, but turning it on will make a huge impact on your upload speed. This is because the radios on cell towers are hugely powerful compared to the tiny radios integrated into mobile devices. If the tower is yelling at maximum volume, your device may be able to hear it just fine – but the tower may not be able to hear the tiny return whisper coming out of your phone. A booster is akin to climbing on the roof and using a megaphone to yell louder – making it a lot easier for the tower to hear you. And this can vastly improve your uploads.

We've seen uploads go from a "barely can send a text" 50Kbps to a "let's video chat" 3Mbps just by toggling on a booster, even in places where the download speeds show barely any improvement at all.

- **LTE Isn't Always Better:** Even if your phone or hotspot defaults to LTE, try disabling LTE (if you can) to see the impact. On Verizon, you will fall back to 3G – an almost always much worse experience. But AT&T's 4G HSPA+ network is in our experience often actually as fast or even *faster* than AT&T's LTE, particularly in fringe signal areas. You may occasionally experience the same improvement in speeds by going to a slower network on Sprint and T-Mobile too. Trial and error is the only way to know for sure.

We often see weak LTE signals that even after boosting are not as usable as a strong nearby 4G signal.

- **Separation Matters:** Boosters thrive on distance between the inside and outside antennas. The ideal distance between antennas for a 50dB amplifier (the most powerful approved by the FCC for mobile use) is actually 40 feet! That distance is impossible to achieve on most RVs, but it helps to do your best. Because of the need for separation, a lot of more powerful home or small office amplifiers will never be a good fit in an RV – they may just be too strong to avoid oscillation at all, no matter what you do.

- **Aim Away:** To reduce the need for separation, some boosters (particularly models intended for fixed locations) come with directional antennas. Make sure that the outside antenna is pointed well away from the inside antenna to reduce your risks of oscillation and the booster shutting itself down.

91

You can do some experimentation to figure out which direction has the strongest signal to aim your outside antenna towards, or you can use some online tools to look up cellphone tower locations. These are very incomplete lists, but they may help:

- Signal Finder (play.google.com/store/apps/details?id=com.akvelon.signaltracker&hl=en) Android App

- OpenSignal.com

- **Never Boost Without an Antenna:** Boosters are not designed to operate without antennas connected – and on some, doing so will void the warranty and potentially fry the internal circuitry. If you are swapping antennas, *always* power down first!

- **Multiple Boosters Can Lead to Headaches:** It is sometimes handy to have different boosters on board, but be careful trying to use more than one at the same time. Do some experiments to make sure that they are not picking up on each other and actually degrading the signal instead of improving it.

- **Don't Boost When You Don't Need It:** If you have a good signal, a booster isn't going to make it better – and it could actually make it worse by confusing your phone or causing interference on the cellular network. This is less of an issue now with the new generation of FCC-approved boosters, but a pre-2014 booster always left on may have been a help in remote areas but actually cause problems for the network in urban areas. The general rule of thumb is to turn off your booster when you don't need it to avoid potential interference.

- **Turn Off Gadgets You Don't Need to Boost:** If you leave all your tech with their cellular radios on, your booster can end up distracted trying to boost everything nearby at once. We've observed that boosters often perform a lot better if other nearby devices that are not actively in use are put into airplane mode, so that the booster can focus on just one or two key devices. Turning off extraneous gadgets may be especially helpful if you are struggling to keep your booster from oscillating.

- **Beware Crossing Borders:** Early 2015 is seeing boosters just becoming certified for both US and Canada use. If yours is not licensed for use in Canada or Mexico, you do not want to get caught using an unauthorized booster in a foreign country. Check with your manufacturer and purchase one certified for the countries you want to utilize them in.

Boosters licensed for use in the US will have an FCC certification number stamped on them, and boosters licensed for use in Canada with have an IC certification number. Some boosters may have both.

- **If You Get "The Knock" – Shut It Down:** It is not hard for carriers and the FCC to track down a malfunctioning booster that is causing interference with the local cellular network. We have spoken to a handful of RVers who have gotten a knock or letter telling them they need to shut down their booster until they can resolve the interference. You are required to comply with these requests, and the penalties can be severe if you ignore them.

  If you get contacted, shut down your booster immediately. Contact your booster manufacturer or the vendor you purchased it through. In the past, we've heard of examples of free new units being sent out to replace a defective booster.

  Theoretically, the new FCC-certified designs should be much less likely to cause any issues.

Can I Change Around Antennas?

One of the stipulations of the new FCC rules is that consumer boosters can no longer be sold other than as part of a kit that includes all necessary wires and antennas. This is meant to ensure that whatever is installed matches what was submitted to the FCC for testing.

There is nothing that technically prevents an end-user from changing around antennas at a later date, and the rules do allow for booster antenna upgrade kits to be sold. But to stay compliant, all additional antennas should at least meet manufacturer specifications.

In other words, if you want to use a different antenna with a booster, contact the manufacturer for advice and recommendations. We already know that both weBoost (formerly Wilson Electronics) and Maximum Signal intend to offer special RV kits with new antennas to go along with their mobile boosters.

If you choose to use an antenna that has not been officially tested with your booster, try at least to match the basic specifications of similar antennas that have been.

A Signal Too Far

No matter how powerful a booster you have, sometimes you might be able to see a strong signal from a cell tower – but the tower still ignores you.

The timing of cellular signals is so precise that the speed of light intervenes, and if it takes too long for the tower to hear your device's reply, it moves on to listen to other signals.

You'll most likely encounter this in the mountains: If you have clear line of sight to a tower far away, you might seem to have a great signal – but nothing you send will ever go through.

Connections to GSM cell towers are actually impossible beyond 22 miles (35 kilometers) because of the timing required. LTE towers have limits too – but they are configurable and can be dialed up much farther in fringe areas. But if you are beyond the range the network engineer planned for, the tower will always end up ignoring you – booster or not.

Old CDMA phones (Verizon & Sprint – pre-LTE) do not have a built-in distance limit, and 30–45 miles is possible in perfect conditions.

The Simple Trick to Expanded Cellular Booster Coverage

To avoid oscillation, mobile-approved boosters come bundled with interior antennas with a very small circumference of boosted signal. In some cases, if your phone or hotspot is more than 3' away from the interior antenna, it will not benefit much from boosting at all.

If you absolutely want to have the most powerful booster possible with you on the road, offering boosted coverage all throughout your rig and even outside – it is going to be a challenge. If you are up for ignoring the FCC rules, juggling directional antennas, and dealing with the downsides – you could probably make a full-home booster work.

On the other hand, you could use this one simple trick to get inside coverage for phone calls and data, all throughout your rig.

Rather than trying to create a large boosted zone that covers your entire interior, set up a tech cabinet or desk area where you mount the interior antenna to create a small boosted zone – and this is where you should set up your gadget-charging station.

Then use a WiFi hotspot feature or router to enable using your boosted connection everywhere in your rig. One gadget in the tech cabinet acting as the WiFi hotspot will get all your other tech online, whether they are in the boosted zone or not.

For voice, a Bluetooth headset or speakerphone accomplishes the same thing.

If you are the type of person who loves having a handset, you can even get a Bluetooth-enabled phone setup – and place a station on your nightstand and at your desk while your phone stays charging away in the tech cabinet, benefitting from maximum boost.

New FCC Booster Registration Requirements

New rules from the FCC went into effect on May 1, 2014 – outlawing sales of old booster models and paving the way for a new generation of cellular

94

boosters designed to reduce the potential for causing interference to wireless networks.

Waiting for these new standards and the accompanying certification process had essentially frozen the market for cellular boosters, holding back new models for over a year. Consumers wanting boosters with LTE and 4G support have been left with incredibly few choices, frustrating bandwidth-hungry nomads everywhere.

But at last, by mid-2014, new boosters are shipping!

But these boosters all now come with a scary mandated warning label:

> *BEFORE USE, you MUST REGISTER THIS DEVICE with your wireless provider and have your provider's consent. Most wireless providers consent to use of signal boosters. Some providers may not consent to the use of this device on their network. If you are unsure, contact your provider. You MUST operate this device with approved antennas and cables as specified by the manufacturer. Antennas MUST be installed at least 20 cm (8 inches) from any person. You MUST cease operating this device immediately if requested by the FCC or a licensed wireless service provider.*

All the major carriers have already issued blanket consent for the use of the new generation of FCC-approved boosters on their network, so you don't need to ask any of the big four for permission. But that doesn't get you off the hook from registering.

Old boosters that do not support the new network protection features are no longer legal to be sold, though they are still OK to use with some carriers…for now, at least.

But according to the new FCC rules, you are now required to register all old boosters too.

The currently shipping boosters don't come with any instructions on where to go and register – just a warning sticker saying you MUST.

Here are the links to each of the major carrier's registration pages:

- AT&T (www.attsignalbooster.com/)
- Verizon (www.verizonwireless.com/wcms/consumer/register-signal-booster.html)
- Sprint (www.sprint.com/legal/fcc_boosters.html)
- T-Mobile (support.t-mobile.com/docs/DOC-9827)

At the time of this book's publication, the registration systems were still relatively new and rough around the edges. For a more updated and comprehensive list, check our knowledge base at:

www.rvmobileinternet.com/resources/booster-registration

FCC Booster Registration FAQs

Why Register?

The primary purpose of the registration databases being built is to help with network troubleshooting issues. If a defective booster is wreaking havoc on the network, the registration info may help carriers track down and isolate the problem before it causes too much interference.

There really isn't a downside to registering, other than just a little bit of hassle.

What if I Don't Register?

You will not be fined or hauled off to jail. But you might be required to cease and desist if your booster is caught causing any network issues.

This general leniency only applies to consumer-level boosters. If you install a booster labeled "for industrial use" without having documented explicit permission from a carrier, you may be facing "penalties in excess of $100,000."

And if you ignore a request from the FCC or any licensed carrier to stop using a booster that is causing interference…well, then you are just asking for trouble.

Are Old Boosters Still Allowed?

Old boosters have not been grandfathered in – but they have not been banned either.

Here is how Verizon explains it:

> "Verizon also tentatively approves the use of consumer signal boosters that do not meet the new network protection standards. This approval is provided only for the boosters not causing interference and may be revoked if the particular booster or booster model is found to cause interference issues. To help avoid possible interference issues, however, Verizon recommends that customers who need signal boosters replace existing boosters as soon as possible with consumer signal boosters that meet the new network protection standards."

96

How Do I Register as a Mobile Consumer?

All the registration forms request some subset of the following information: owner's name, operator's name (if different), contact phone number, booster make, model, and serial number, date of initial operation, and *installed location*.

Some of the forms ask whether the booster will be mobile or installed at a fixed location, but many of them seem to not have considered mobile users – especially mobile users without a fixed-location home base.

In those cases, we recommend using your mailing address.

What if I Have Multiple Devices on Multiple Networks?

The guidance from the FCC says that you should register with every carrier where you will regularly be connected. You need to register once per booster per carrier – it does not matter how many devices you are connecting.

What About Friends Who Use My Booster? Guests?

The FCC has ruled that it is perfectly fine for friends and visitors to take advantage of your booster without explicitly registering. But if you have housemates who are making regular use of your booster, they should register with their carrier too.

What About WiFi "Boosters"?

These new rules only apply to cellular boosters, not WiFi repeating systems. So you do not need to do anything about products you might be using for your WiFi signal enhancing – such as from WiFiRanger, Alfa, JefaTech, Ubiquiti, Wirie, etc.

If you're using something from companies like weBoost (formerly Wilson Electronics), TopSignal, SureCall, Maximum Signal – these are cellular boosters and the new rules apply.

Boosting Summary: Keep Realistic Expectations

A lot of people invest big dollars into cellular boosters and WiFi repeaters expecting miracles, and many end up being disappointed when their lofty expectations are not met.

Boosting can only do so much. If there's nothing to boost, no amount of signal amplification can make something out of nothing.

And even if you get a stronger signal thanks to a booster – if the real speed bottleneck is located upstream, then you might not actually see any practical improvement anyway, at least when it comes to your online experience.

But if you keep realistic expectations, a cellular booster and a WiFi repeater can grow to become some of the most essential elements of your tech arsenal.

We've lost count of how many places we've been that would not have been doable without a booster on board.

When it works, it becomes absolutely indispensable.

Antenna Selection & Installation

by guest author Jack Mayer

Antennas are key to successful mobile internet – for both WiFi and cellular.

The antenna you select, or that the manufacturer selects, has more influence on your internet experience than almost any other single piece of your equipment. But for most people, antennas are mysterious devices built into their tech or sitting forgotten on their roof.

How do you tell if the antenna you are using is the optimal choice?

In this chapter, we take a look at antenna types, some of the terms and technology used, how to select an appropriate antenna, and how to mount it once you own it.

Although antennas are an important part of any connectivity solution, they are dependent on clear line of sight, or near clear line of sight, to the signal source.

Getting the antenna up high, away from obstructions, is just as important as the performance characteristics of the antenna itself.

Antennas are an incredibly complex topic if you want to understand all the intricacies of them. People earn PhDs in antenna theory and design. Here we look at the practical aspects of antennas – not antenna theory. We gloss over or take some liberties with some of the intricacies of antenna technology to keep it simple and less boring.

Antenna Performance & Gain

The primary performance attributes used to describe all antennas and amplifiers is gain.

So what is gain, and how does it affect your antenna choice?

99

Gain is a measurement of the ability of an antenna to focus power on a target location, or the increase in the amount of power brought about by passing a signal through a booster.

If you think of a traditional bare light bulb, the light emanates in all directions. If you then think of a flashlight, that same bulb has the light focused with a reflector so that it is more directional, and thus appears brighter. And if you increase the wattage of the bulb in the flashlight so that you can see it from even farther away, that is amplification and represents a further gain in transmission power.

In this flashlight example, the overall gain would be the amount the perceived brightness changed – a combination of better directionality provided by an antenna and increased power by a boosting amplifier.

Gain is always measured relative to something. In an antenna, the measure of gain is sometimes expressed in dBi . This is decibels relative to a specific type of theoretical reference antenna – an isotropic radiator - that radiates equally in all directions.

A higher gain means that more power reaches the end point of the transmission.

You might also see gain specified as dB (not dBi). Gain measured as dB is always less than the dBi measurements, and is a more realistic measure of relative performance. Figure on around a 2 dB loss when converting from dBi to dB. In other words if a manufacturer specifies an 8 dBi gain on an antenna, expect an actual gain of around 6 dB in use.

Make sure that when comparing specifications on antennas that you are comparing "apples to apples" (dB to dB, not dBi to dB).

Gain is logarithmic. Without getting into the math, the simple rule of thumb is that for each 3dB of gain, the measured power at the receiver is doubled.

Gain in an antenna is often highly misunderstood – but it is simply the amount of focusing of the signal. A 0 dBi gain antenna radiates energy in all directions equally, while a 5 dBi gain antenna focuses the energy in certain directions more than others – but the total amount of energy radiated remains the same.

The gain of a given antenna is measured at the point of the peak energy concentration.

Related to gain is beam width. Beam width is a measure of the amount of focus the radio signal has. The narrower the beam width is, the more focused the antenna and the higher the gain.

Beam widths are generally stated in degrees of horizontal and vertical focus. The horizontal beam width describes where the signal radiates outward from the antenna. An omnidirectional antenna has a 360-degree horizontal beam, since it is radiating in all directions around it. However, it will have a vertical beam width that varies. The vertical beam width describes the angle of the signal radiating outward, relative to the ground.

To keep things simple and more understandable, we are going to take the liberty of categorizing the typical antennas used for cellular and WiFi into two major categories based on their signal radiation characteristics: omnidirectional (omni, for short) and directional.

With an omnidirectional antenna, a higher dBi rating generally means a tighter vertical beam, since the energy gets focused more in a plane parallel to the ground.

For example, an 8dBi omni antenna may have a 15-degree vertical beam, while a 15 dBi omni may have a narrower 6-degree vertical beam.

The pattern from an omni antenna can generally be thought of in terms of a donut shape, where the antenna is the center of the donut and the shape of the donut represents the signal radiation. The signal is focused outward from the antenna in the horizontal plane – so directly above and below the antenna there is (relatively) little signal.

Thus a lower gain antenna may actually outperform a high-gain antenna if the signal is bouncing among hills, mountains, or city structures – or if the tower is located very high relative to the receiver.

A high-gain omnidirectional antenna, on the other hand, is most useful when towers and receivers have a clear line of sight to each other parallel to the ground – in other words, when the tower is farther away and on the horizon.

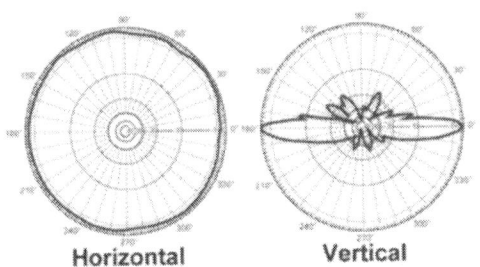

Horizontal   Vertical

This diagram shows an omnidirectional antenna radiation pattern. In the horizontal plane – looking down on the antenna from above (the antenna is in the center) the antenna radiates equivalently in 360-degrees.

The vertical diagram represents a "side view" of the omnidirectional signal. You can see that there are some "lobes" of signal that radiate upward, and others (smaller) that radiate downward. But most of the signal is sent directly away from the antenna parallel to the ground.

101

The vertical beam is important in omnidirectional antennas – higher dBi omnis have very tight vertical beams, so if they are used close to a receiving station they may "miss" with the main part of the signal.

Also - because the signal travels parallel to the ground, it is important for omnidirectional antennas to be mounted vertically. While it might not need to be specifically aimed, an omnidirectional antenna mounted flat will deliver very disappointing results.

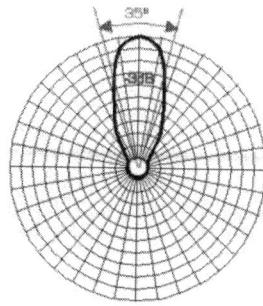

This diagram is a view of the signal from a directional antenna, looking down on the antenna from above. The signal radiates outward from the antenna in the center – the beam width in this case is 35 degrees – a sharp contrast to the omnidirectional antenna above.

You can see that the signal is focused forward in a single direction – making aiming critical.

Choosing an Antenna

Using the best available equipment and directional antennas, you can expect to get usable WiFi at up to a half to a full mile away from a quality base station, assuming a clear line of sight to the station.

In some circumstances – like across water – you can capture WiFi from much farther (as much as two+ miles), but typically the limitation is the power of the access point. WiFi hotspots (access points) are not typically designed for long-distance networks, so the hotspot may not have enough signal to reach your location, even though you have enough power to reach it.

Cellular is very dependent on how the tower you are accessing is configured – some towers have distance constraints engineered into them. It also depends on the frequencies and protocols being used. But it is fair to say that 18–20 miles is generally the best you will see for cellular communication, no matter what antenna and booster you use.

Every antenna is designed and optimized to operate over given frequency ranges.

For WiFi, there are two frequency bands: the 2.4GHz and 5.8GHz ranges.

Cellular frequencies, on the other hand, vary by carrier and location (see the chapter "Understanding Cellular Frequencies"), so cellular antennas are challenged to pick up a wider range of frequencies than are WiFi antennas – and are often only optimized for a few.

102

When looking at cellular antennas, you must make certain that the antenna is designed for the frequency bands your carrier is using.

Generally cellular and WiFi antennas are not interchangeable.

Antennas should have a spec sheet that lists the gain at various different frequencies – use this to confirm how well the antenna will perform with your carriers.

Antennas also come in various types or form factors.

Most people are familiar with the antennas that are seen mounted on vehicles. In the old days these were simple metal shafts. But for high-performance computer and cellular applications, there are more specialized antennas.

The two basic types of antennas – omnidirectional and directional – are available supporting both cellular and WiFi frequencies.

Omnidirectional Antennas

These are the easiest antennas to use, since they require no aiming. But radiating energy in 360 degrees has the disadvantage of diluting the signal. The receiving station – be it a WiFi access point or a cellular tower – is in a fixed location, thus most of the signal energy is wasted in transit to it.

Omnidirectional antennas are also more prone to outside RF (radio frequency) interference – they are, after all, receiving both signal and noise from all directions.

For an antenna in motion, omnidirectional capability is a requirement. You have no idea where the receiving station is so you have to broadcast in all directions in order to ensure that you can hit the receiver.

Most mobile users will benefit from the passive ease of omnidirectional antennas. If you pull into an RV park, you will want to access the RV park's access point (AP) to receive WiFi without needing to aim an external antenna. An omni antenna does this quite well. But because some of the energy is wasted by not being focused on the access point, performance may suffer.

The same thing applies to cellular voice and data reception – you typically want access without needing to aim an antenna first.

There are many form factors of omni antennas. Fixed-mount antennas use a bracket mounted to the side of an RV, vehicle, or building, or they may actually be mounted through the roof of a vehicle.

Magnetic mount antennas use a magnet to hold the antenna to a metal surface – typically the roof of a vehicle. But any metal surface can be used – for example, a metal plate held to the roof of an RV with caulk.

Glass-mounted omnidirectional antennas offer a mounting method that can place the antenna high on a vehicle and not damage the body. Glass-mount antennas pass the signal through the glass without needing a hole: The interior patch with the lead is placed directly under the exterior patch that holds the actual antenna. The glass acts as a capacitor, but allows the AC energy of the signal to pass through to the antenna. It is critical for the glass to be nonmetallic – otherwise, the signal cannot be passed. Solar glass or glass with a defrosting grid in it can cause signal issues, so you have to be sure of the type of glass you are dealing with. On an RV, dual-pane glass on windows will not work properly with this type of antenna.

Glass-mount antennas are commonly used on vehicles for cellular frequencies. They can work well if properly installed and tuned. For use on an RV, a higher performance antenna is often preferred, but a glass-mount antenna on the windshield of a motor home can perform adequately if no other option is available.

Some multi-band antennas combine multiple internal antennas into a single casing - with a separate antenna wire for each device to be hooked up. This is the only way that a single antenna should be used to connect to multiple devices - an antenna wire should never be used with a splitter to plug into multiple devices.

Examples of Omnidirectional Antennas

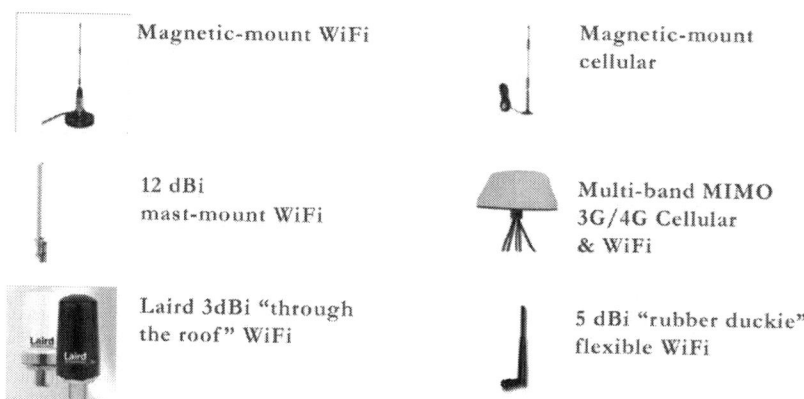

Magnetic-mount WiFi

Magnetic-mount cellular

12 dBi mast-mount WiFi

Multi-band MIMO 3G/4G Cellular & WiFi

Laird 3dBi "through the roof" WiFi

5 dBi "rubber duckie" flexible WiFi

Directional Antennas

With a directional antenna, all of the radio energy is focused into a single direction, with the horizontal beam width generally 120 degrees or less. The vertical beam will also be more focused.

How tight the beam is focused varies by antenna – but it can be as narrow as 10 degrees, or even less.

If you think of an omnidirectional antenna as a lantern, you can imagine a directional antenna as a flashlight. The more powerful the antenna, the tighter the beam is focused, and the more critical aim becomes.

Directional antennas used for cellular and WiFi fall into broad categories.

The widest beam antennas are generally called sectional antennas. These offer wide coverage areas – generally around 120 degrees, so three sectional antennas on a mast working together can cover 360 degrees, and together offer more focused coverage than a single omni antenna could. This setup is commonly seen on cell towers. It is also used for WiFi where only part of an RV park needs coverage.

Next in coverage is a panel antenna. The panel antenna is more focused than the sectional and is available in a range of beam widths as tight as 20 degrees. Panel antennas are often typically smaller than the sectional antennas and are thus a better selection for most mobile requirements.

Yagi antennas perform similarly to panel antennas, with a tightly focused beam width in both the horizontal and vertical planes. Some Yagi antennas have wider beams and require less precision to aim.

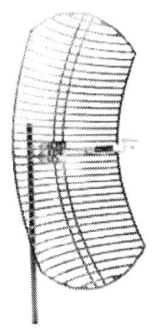

Grid antennas and dish antennas use a parabolic reflector to focus the energy even tighter – generally around 7–10 degrees both horizontally and vertically. There is a feed antenna suspended in front of the grid or dish whose purpose is to transmit/receive signal. The dish acts as a reflector and focuses the signal into a very tight cone. Dish and grid antennas are most commonly used for longer-distance point-to-point connections.

There is some overlap in all of these specifications, but in general the sector antennas have the broadest beam, and the antennas using a parabolic reflector (grid and dish) have the most focused beam.

Grid and dish antennas are much harder to aim than panel antennas.

Understanding Ground Planes

A ground plane is a reflective metal surface that the signal from an antenna bounces off of to better meet the antenna's design goals.

Antennas requiring ground planes are called ground-plane dependent, and having an appropriate ground plane can greatly affect performance. A dependent antenna can become effectively useless without one. You should always ensure that any antenna requiring an external ground plane – typically magnetic-mount antennas – is placed upon an effective reflective surface.

Antennas that do not require an external ground plane are called ground-plane independent antennas. These antennas have the means to reflect and focus the signal without the external ground plane being present. For most ground-plane independent antennas, having a ground plane present will still somewhat enhance signal gain, but it is not required.

The minimum size of a metal ground plane should be one-fourth of the wavelength of the radio signal being broadcast. For most cellular and WiFi transmission frequencies, a disc of some sort of metal at least 8" in diameter is generally considered sufficient and actually provides some room for error.

Most people think that because the typical ground-plane-dependent antenna has a magnetic mount that a nonmagnetic metal will not work as a ground plane. That is not true – aluminum and other metals do work fine. The shape is also not critical: A ground plane can be rectangular or circular, as long as it meets the minimum size measure in all directions. Flat cookie sheets of the correct size, pizza pans, old circular saw blades, paint can lids, and many other things can all make good ground planes for most antennas.

And, of course, metal car roofs work wonderfully – just keep the antenna at least 4" away from vents and sunroofs.

MIMO Antenna Technology

Multiple-input multiple-output (MIMO) technology uses multiple antennas at both the transmitter and receiver to greatly improve both performance and reliability.

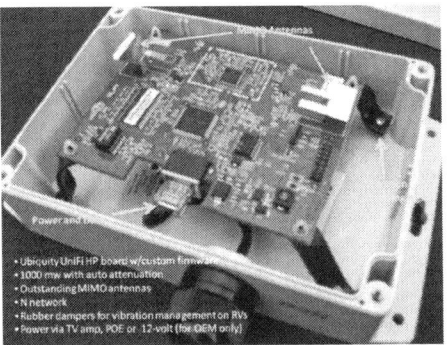

Shown here is the interior of a WiFiRanger Sky combination WiFi router/CPE that has a pair of MIMO antennas.

MIMO is part of the LTE cellular standard and the 802.11n WiFi standard – all "n" networks must support MIMO.

Some LTE devices (like the Pantech MHS291L on Verizon and Netgear Unite Pro on AT&T) support dual external antennas, allowing for the combination of MIMO capabilities with the benefits of external antennas.

Highly Regarded Antenna Sources

- 3G Store (www.3gstore.com)

- Powerful Signal (www.powerfulsignal.com)

- WPSAntennas (www.wpsantennas.com)

Antenna Cables & Calculating Overall System Gain

Another factor that affects performance of all antennas is signal attenuation due to the cable length, size, and quality between the antenna and the radio.

Attenuation is a general term meaning reduction in the strength of a signal over distance. In the context of antennas used in a mobile environment, it is almost always the result of cable length.

You need to use good quality antenna cables and good connectors or the signal will be compromised, defeating entirely the benefits from using a high-quality external antenna or booster.

The lower the frequency of the signal, the less attenuation there is in a given length of cable. So the cellular LTE 700MHz frequencies show significantly less signal loss over the same length of wire than the 1900MHz PCS or AWS 1700/2100MHz frequencies.

WiFi signals at 2.4GHz degrade rapidly over antenna cables, and 5GHz WiFi is not well suited for even short wire runs between the receiver and the antenna. This is why it often makes a lot of sense to put the entire WiFi radio up on the roof and connect to it digitally over Ethernet, which does not suffer attenuation with distance.

Cable attenuation is expressed in decibels of loss – so it is typical to see a cable extension specified as "attenuates signal 2dB" (if it is expressed at all).

Most cellular antennas come with RG-58 or RG-174 cable, with a 12–15' long section often hardwired to the antenna. These wire types are thin and flexible, but for extension cables and longer runs, heftier cables should be used. Here are the cable types you are most likely to see in use in an RV:

Cable Type	Attenuation of a 750MHz Signal	Attenuation of a 2.4GHz Signal	Comment
RG-174	23.6dB/100ft	75dB/100ft	Very thin and flexible but even higher loss than RG-58. Common on magnetic roof antennas and interior antennas.
RG-58	13.1dB/100ft	32.2dB/100ft	Commonly comes on trucker and marine antennas. Suitable for extensions of 20' or less.
LMR240	6.9dB/100ft	12.9dB/100ft	An excellent choice for extension runs inside an RV or up an antenna mast.
RG-6	5.6dB/100ft	N/A	Commonly used for cable TV and satellite wires, but due to the 75 ohm impedance should NOT be used for cellular or WiFi antennas unless you are certain your equipment is compatible (rare).
LMR400	3.5dB/100ft	6.8dB/100ft	Appropriate for small buildings, only rarely used in RV applications. It is a very heavy cable, and does not flex or bend well.

In general, any cable run over 30' needs to be looked at very critically to manage signal attenuation.

Remember that the dB scale is logarithmic – 6dB of attenuation over a wire run will result in half the power getting through as a 3dB attenuation run.

Every connection point adds roughly 0.5dB attenuation per junction – so it is always better to use a single long cable than to chain two or more cables together.

You should never ever use a splitter with cellular and WiFi antennas - splitting a signal will cut the power received in half, and two transmitters connected to one antenna is a recipe for trouble.

Your goal for any extension cable should be a total attenuation loss of less than 4dB over the distance you want to cover.

Low-loss cables are more expensive and are stiff and harder to work with – all things to keep in mind when planning your antenna cable runs.

To determine your overall system gain at a given frequency, you add the gains provided by the outside (and inside) antennas, subtract the attenuation from the antenna cables and extensions, and add any gain provided by an amplifying booster.

An Antenna Evaluation Case Study

Say we want to mount a typical magnetic mount omnidirectional cellular antenna on our RV roof specifically to improve Verizon 4G cellular data performance with a Jetpack modem.

We know in most areas that Verizon primarily operates 4G/LTE in the 700MHz band, although over time they are deploying LTE into all of their bands too – 850MHz, 1900MHz, and 1700/2100MHz.

The Wilson 301103 magnetic-mount antenna spec sheet reveals it will provide a gain of 1.9 dBi in the 700MHz frequency range, and includes the RG-174 ten-foot cable and connector.

That is not much gain in the 700 MHz band, and if we add an additional 10' RG-58 extension cable, we attenuate the signal an additional 1.3dB, plus 0.5dB for the connector. This leaves us with 1.8dB line loss – the gain from the antenna is negated by the extension!

But let's look at this same antenna in the 850/1900MHz band that Verizon uses primarily for voice and 3G data. At 1900MHz, it has a gain of 6.12dBi, and at 850MHz a gain of 5.12 dBi, so we know that for voice calls this antenna will work well, even with an extension cable.

This antenna was clearly designed as a dual-band 850/1900MHz antenna, and only marginally performs in the 700MHz bands. In other words, this is a bad antenna choice for LTE and mobile data!

What would be a better antenna in this situation? Next, let's look at the popular Wilson Trucker antenna, 311133. In the 700MHz bands its gain is 3.1dBi; at 850MHz, 4.1dBi; and at 1900MHz, 5.1dBi. So it performs better across the Verizon bands we are interested in, and would be a better long-term solution for our 4G data example.

Need even more performance? You would likely add an amplifier next, and that would improve gain quite a bit. For example, let's look at the Wilson Sleek 4G 460107 cradle-style booster.

Again, focused on boosting 4G data, we insert the Jetpack hotspot into the cradle and use the provided stubby omnidirectional antenna. This antenna only has about 1–2dBi of gain, so it is not very powerful. But the amp provides a max 23dBi of gain. So even with the low-performance antenna provided with the amp, this combination will outperform the Jetpack alone in combination with the Trucker antenna. If you plug the Trucker antenna into the Sleek, you will further maximize your performance potential.

From the above examples, you can see that performance is dependent on a variety of factors that interrelate. Varying only one factor may not provide the maximum available performance in a given situation.

You must take all factors into account when designing your system.

Measuring Antenna Performance

Getting a valid measurements to evaluate antenna performance can be tricky without specialized gear.

There are many variables that affect performance, and some of them are out of your control. But the bottom line is:

> **For Cellular** – Are your voice calls going through without issue? What is your data speed as tested with a tool like speedtest.net, or speedof.me? The actual results are what count more than anything else.

> **For WiFi** – As above, what is your data speed using the speed-testing tools? Over what range are you able to remain connected?

While bottom-line results are what count, you can also view the raw connection strength to your cellular tower or to your WiFi access point with most phones, modems, and routers.

They all vary in how they provide this measure. What you are looking for is something more granular than just signal bars – look for the signal RSSI or the raw measurement of received signal power in dBm.

If you change antennas, move antennas, add extension cables, or add amplification, these numbers when comparing before and after measurements can tell if your signal improves.

110

If it does improve, you would typically have better quality voice or faster data speeds.

On the Roof – Mounting Methods & Examples

We know that for both WiFi and cellular signals the main culprits that affect your results are:

- Poor line of sight (LOS) to the source signal (cellular tower or WiFi access point)

- Radio equipment that is not powerful enough, including poor antenna choices.

Mounting your high-quality antenna – for either WiFi or cellular use – on the roof of your RV or vehicle helps to overcome poor LOS (line of sight).

Getting the antenna above obstructions such as other RVs will greatly improve your overall performance. Just doing this often improves the signal enough so that amplifiers or other signal-boosting techniques are not required.

Let's look at some common mounting methods for both cellular and WiFi antennas.

Interior Mounts

Though the altitude that comes from roof mounting is ideal, sometimes it is not practical. In those cases - with a little trial and error at each new stop you can find an optimal interior window to place your antenna inside of.

This photo is of a cellular panel antenna with a temporary mount placed on a motor home dash. You can see a Wilson cellular amp sitting next to it.

Although performance would be improved by mounting this panel antenna above the roofline of the motor home, this provides sufficient gain in many circumstances. It does have to be pointed toward the cellular tower to help, though.

111

Side Mounting

Permanent or temporary side mounting of larger antennas on vehicles works well. Especially on RVs, side mounting on the driver's side offers some protection from tree limbs and low bridges that might affect an antenna mounted flat on the roof.

This picture shows a permanently mounted Wilson Trucker cellular antenna with a flexible spring base mounted next to a temporarily mounted WiFi antenna that is a replacement for a WiFi router's built in "rubber ducky" antenna.

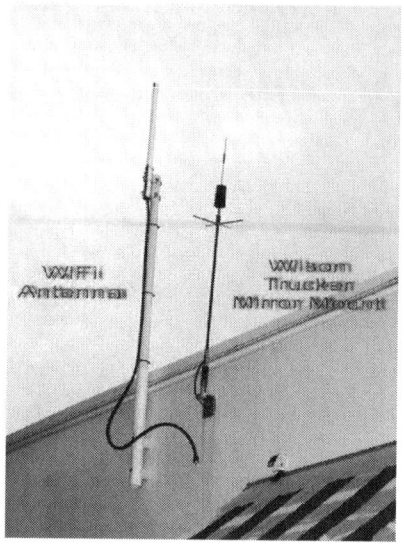

On this RV, the WiFi antenna is dismounted and snapped into a carrying bracket for travel – you can see the bracket just above the window awning. The antenna mast is made from a piece of PVC tubing and screwed to the side of the RV.

The obvious downside to this method is that you have holes in the side of the RV. However, holes in the side are often more desirable than holes in the roof, since they tend to be less prone to leaks.

From an appearance perspective, however, side mounting may not be desirable.

Roof Mounting

The key challenge with roof mounted antennas is protecting them from damage while underway.

This is commonly accomplished by using shorter height antennas, or taller antennas that can be retracted flat.

Here is a Wilson low-profile cellular antenna attached to the side of a solar panel mount with a piece of aluminum plate.

This mounting location allows the antenna to function while traveling, but has the disadvantage of potential shading of the panel. This is not a magnetic-mount antenna, but is an NMO (new Motorola) style.

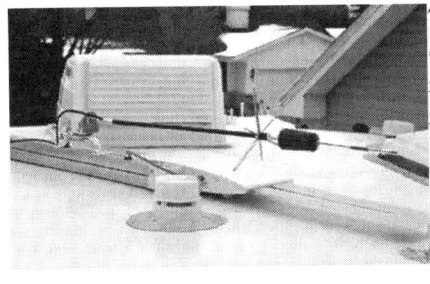

This pictures shows a Wilson Trucker cellular antenna permanently mounted to a batwing TV antenna. It is simply attached to a stainless steel handle screwed to the batwing base. The normal mirror mount is used.

The disadvantage of this method is that the antenna is not available for use when driving because the antenna is lowered. The advantage of this method is that the antenna is out of the way on the roof and is up high when the batwing is deployed.

It does not interfere with normal use of the TV antenna.

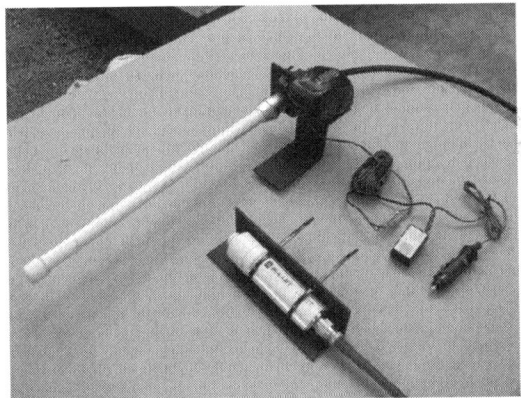

Pictured above is a motor-drive mount that lifts an 8dBi WiFi omni antenna used with a Ubiquiti Bullet as a CPE. The Bullet is mounted separately, flat on the roof. The motor drive is 12 volt and is used to raise and lower the antenna much like the TV batwing does.

It can be mounted on a vehicle or on an RV roof. It is attached to the roof with VHB tape, or 3M 5300 adhesive. No roof penetration is required. There is a switch to control it mounted in the interior of the vehicle.

Roof mounted antennas should also be kept to the left side of the roof, if possible - lessening the risk of being hit by a low branch on the curb side of the rig.

113

Mast / Flagpole Mounting

Mast mounted antennas have the advantage of height, but the disadvantage of setup time. They often work best when used to compliment a roof mounted antenna only when needed, not as a primary antenna.

In these photos we see two temporary mounts using a 16' collapsible painter's pole repurposed as a mast. This pole reduces down to less than 8' so it stores fine in most RVs with coach width storage space. There are smaller poles available as well. The picture on the left is a Wilson Trucker mounted to the top of the pole with electrical tie-wraps.

The antenna line goes thought a window in the slide or is shoved around the slide seal. This method gets the antenna quite high. In this particular location, the cell tower was 18 miles away, and there was no detectible signal at the RV. With use of the antenna alone, there was a barely detectible signal. The addition of an amplifier gave solid voice and 3G data.

The painter's pole is attached to the ladder with tie wraps. Some people permanently attach the painter's pole to the ladder and simply lower the top section when moving locations. The extra cable is looped around the ladder for travel.

The second picture is a WiFi CPE attached to the painter's pole with a clamp mount. The top part is an 8dBi omni antenna, and the bottom part is a Ubiquiti Bullet.

Again, this is a temporary mount and is attached to the ladder in the same fashion. The device is powered by Power Over Ethernet (POE) and is used to improve long-range WiFi capture.

114

Another way to gain some altitude is via a collapsable flagpole – providing a way to both fly your colors and enhance your signal. There are many RV flagpole kits available – the FlagPole Buddy mount (www.flagpolebuddy.com) is particularly impressive and provides an easy way to quickly raise and lower a flagpole.

The FlagPole Buddy comes in sizes ranging from 12' to 22' high.

It is an easy thing to strap a WiFI CPE like the Ubiquiti NanoStation or a WiFiRanger Elite to the top of a flagpole, getting above all nearby obstructions.

If you feel a need to get even more altitude than a flagpole can provide, a retractable mast from a TV news truck can get your antennas over 60' into the air – if you really want to get crazy.

One note on using a flagpole: you want to make sure that you do not attenuate the signal too much with extra cable length, which could easily negate any benefit from placing the antenna higher up.

Because of this, in most cases very high mast mounting is best used with equipment that uses Ethernet to carry the signal - which means WiFi capture and not cellular.

Using a cellular antenna at the top of a tall mast requires careful planning and a heavier cable than just roof mounting.

Directional Antennas & Aiming

Mounting directional antennas requires the ability to point them towards the signal source.

Shown above is a CPE with a directional panel antenna used to capture WiFi. This is permanently mounted to the RV ladder with PVC schedule 40 pipe. The pipe structure is built with a 1.5" pipe inside a 2" pipe. The inner tube is longer than the outer tube and is used to extend the CPE above the roofline when not traveling. The inner pipe has 8 holes in it and is rotated to point towards the signal source. The outer sleeve has two holes. The inner holes are aligned to the outer holes and pinned in place to lock in a given direction.

For travel, the inner pipe is lowered so that the CPE is below the top of the ladder, and it is pinned in place securely.

The required Ethernet cable is routed to a jack on the side of the RV. This mounting method can be used for any directional antenna.

116

A directional antenna can also be mounted to the batwing antenna. Here we see two versions of a WiFiRanger Mobile unit mounted to the TV batwing. Although these both use an omnidirectional antenna (a Laird stubby and an 8dBi omni), you could also mount a directional Ubiquiti NanoStation in the identical fashion.

The NanoStation uses a directional antenna and performs at a far higher level than the mobile unit using the Laird antenna.

The con of mounting a directional antenna using the batwing is that the TV cannot use the batwing for aiming at the same time, since the antenna is dedicated to pointing at the WiFi or cellular source.

Pictured below is a directional CPE simply screwed to the top of a batwing. This is a very simple but effective mounting technique. Power is supplied over Ethernet, and there are no issues with cable length as there are with analog antennas.

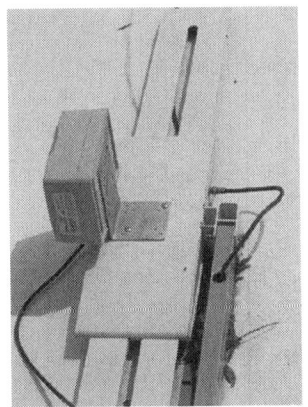 

Getting Creative

You can fabricate antenna mounts from most anything to creatively solve mounting challenges.

PVC works well in a lot of cases and has the added benefits of not rusting, being readily available, and easy to work with.

For example - the WirENG BoatAnt cellular antenna shown here is intended to be mounted to a pole, and has an integrated mounting plate. But it can be roof mounted by adapting some schedule 40 PVC piping and using caulk - as shown.

117

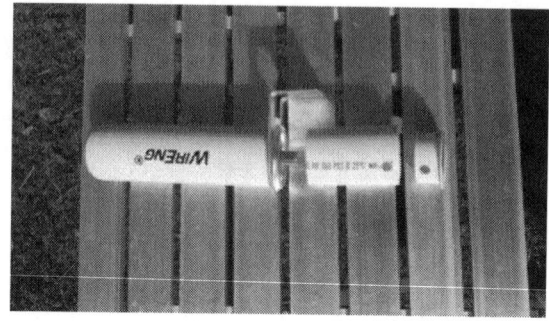

A magnetic-mount antenna can be installed on any RV roof – even on fiberglass. You simply have to provide your own metallic plate to act as the ground plane.

Ground-plane size depends on the frequency of the transmitted signal. For the cellular and WiFi frequencies and for smaller magnetic-mount antennas like the one pictured here, a 4" diameter plate is usually sufficient.

Most magnetic antenna manufacturers will specify the size required.

To be absolutely sure, an 8" diameter plate always works. Shown in this photo is a 12"x12" plate – this is overkill, but doesn't hurt.

The plate does not have to be heavy/thick and it only needs to be magnetic if you want to use the magnet on the antenna base to attach the antenna.

To attach a magnetic antenna to a fiberglass or rubber roof, take your metal plate and spray it with your choice of rust-resistant paint, and then simply attach it to your roof with a compatible caulk.

Then place the antenna in the center of the plate, and let the magnet hold it.

And then at the side of the plate, secure the wire to the roof with a puddle of caulk.

The advantage of this method of attachment is that if a tree limb hits the antenna, it will simply flip off the mounting plate and be retained on the roof by the caulk puddle. Later, it is a simple matter to right the antenna back on to the plate.

A rigidly mounted antenna on the other hand is easily damaged by tree limbs.

Cable Routing & Management

The biggest issue most people have when mounting equipment to their RV is the fear of penetrating their RV roof. There are various ways to do this with little issue.

The best way, if it is possible, is to drill a fairly large hole in the roof and run a conduit from the communications cabinet directly to the roof. Then use caulk to mount an electrical utility junction box over the conduit. This provides an easy way to route cables as required – and the conduit makes it easy to change or add cables later.

Even if not using conduit to enter the RV, use a junction box to cover the entire hole in the roof. This provides the most weather-resistant seal possible.

The wires exit the junction box on the roof via a weatherproof cable connector. These connectors may be found at any Home Depot or Lowe's in the electrical section. Shown here is the communications box with a Category 5 Ethernet cable and an RG-58 cellular antenna cable exiting it. On the right is the solar combiner box.

Both of these have cables exiting the bottom of the box and entering the RV directly through the roof. For the wires exiting to the RV roof, try to face the cable exits to the rear of the RV, where possible. This helps keep water from being driven into the box by wind as you drive. If you cannot exit from the rear, then use a side exit. Avoid a front exit.

Self-leveling caulk compatible with the roof type is used to hold the box in position and simultaneously seal it. No screws are needed or used.

Wires can be held in place on the surface of the roof by pools of caulk – make sure the caulk is compatible with the roof material.

Place a puddle of caulk on the roof and lay the wire through it. Use tape on either side of the puddle of caulk to ensure the wire stays pressed into the puddle. When dry, remove the tape and place some additional caulk over the wire and the original puddle of caulk. This will ensure the wire will not

119

pull out of the caulk. Use your best judgment about the distance between puddles – four to five feet is usually close enough.

If you have a number of wires running together, it is generally acceptable to embed two of them in the caulk puddles and then simply wire-tie the others to those two. They will not go anywhere. Make sure you use UV-resistant wire ties. Even these will degrade in a few years, so keep your eye on them when doing your roof maintenance.

Another retention method that works well is to use a small piece of Eternabond sealing tape over the wire. A two inch long piece is adequate place them every 4 feet or so. Generally, the caulk method is preferred since it is far easier to remove, if required.

Also, make sure that any wire on the roof is UV resistant. Antenna wires are fine to expose to the weather. Ethernet cable needs to be outdoor-rated cable. Do not use plenum-rated Ethernet cable if you are wiring a device yourself.

An alternative to making a new hole in your roof is to take advantage of an existing vent opening. The refrigerator roof vent is often a convenient place to run wires down from antennas.

Any time you run a wire, think ahead. What if you ever need to upgrade or replace it? Using conduits makes future wire runs and antenna upgrades potentially a lot easier.

120

Satellite Internet

Before cellular internet and prevalent WiFi hotspots became the norm, satellite internet was the ultimate option for getting online at better than dial-up speeds while mobile.

Back then, if you walked around any high-end campground, the signature blue glow of Datastorm dishes mounted on roof tops meant one thing: the people inside were online, while everyone else made do with dial-up or nothing.

But sadly, the glory days of relatively affordable, available-almost-everywhere satellite broadband are now fading into the past for mobile users.

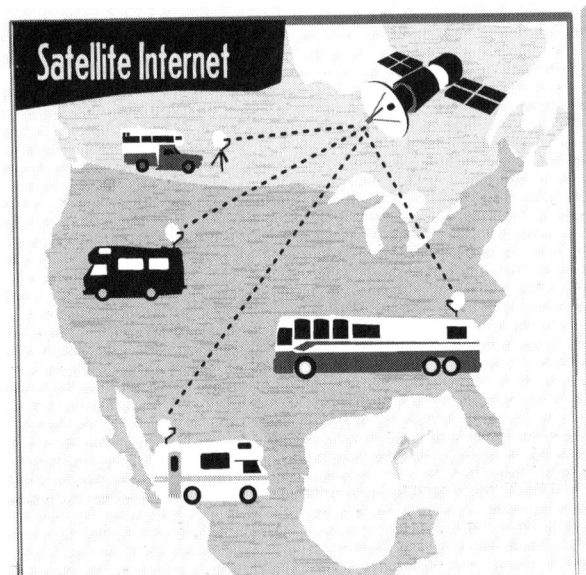

Satellite Internet

Quick Glance

Pros

Signal almost everywhere

Unlimited overnight

Cons

Latency

Bandwidth caps

Slower

Cheap and increasingly widespread cellular 3G and 4G has eaten into the advantages of satellite, so now fewer and fewer customers are willing to put up with the costs and considerable headaches associated with orbital internet.

And with fewer customers to serve, satellite companies are either going out of business or finding more lucrative markets than budget-conscious (in comparison to big companies and government entities) RVers.

But even though many longtime satellite customers are ditching their dishes as the cellular coverage maps increase, it's still a great option for some situations.

Especially those situations where it is the ONLY option.

If you're planning a good amount of time off the coverage map or planning to cross borders into Mexico and Canada, satellite will give you the best odds of getting online – as long as you can point your dish to your assigned satellite in the southern sky.

And if you are willing to redefine what it means to be online, you might find that you can get creatively connected without needing a dish and a traditional satellite broadband plan.

Satellite TV and Internet Are Not the Same!

You'll often see satellite TV providers Dish Network and DirectTV advertise bundled packages that includes internet service – but this is almost never satellite-provided internet.

These plans are intended for stationary satellite TV consumers to combine their TV, internet, and phone bills into one. The satellite TV provider contracts out to local DSL or even cable companies to provide the actual internet service – requiring a hard-wired connection. Even when they do offer actual satellite data plans (like Dish's DishNet), these are provided only to offer service to rural customers outside the range of cable and DSL, and they are strictly for fixed-location installs only.

In other words – not mobile friendly at all.

TV satellite dishes are not two-way devices, and a two-way capability is required for satellite internet to send and receive data. TV dishes are built to receive only and cannot be used for internet services.

However, some satellite internet dishes can be set up to receive TV signals by means of a "bird-on-a-wire" (aka BOW) kit that tacks on a TV receiver to a data dish. At least this way you can avoid needing multiple receivers on your roof.

122

The Signal: Spot Beam vs. Broad Area

Traditionally, communication satellites broadcast with an antenna that could reach an entire continent.

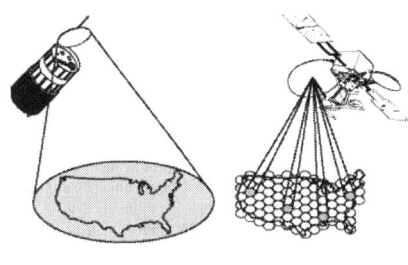

This is great for mobile users – the satellite doesn't know or care whether you are in Boise or Boston, in the Black Rock desert or back-country Georgia.

But it is also horribly inefficient – with every user assigned to a particular satellite and channel having to share the signal.

Newer satellites use spot-beam technology to cover the nation with many small focused signals rather than a single blanket signal – allowing for many more users to communicate at once.

This allows for cheaper service and faster rates. But it also means that if you travel more than about 100 miles from your "home" address, your satellite service will no longer work at all – assuming you can even aim the dish correctly, since spot beam receivers are even more sensitive to needing absolute precise aim!

In other words, spot-beam satellite is at best useful for nomads who plan to travel in very tight circles or will be moving locations very infrequently. Personally, we consider it incompatible for mobile lifestyles.

If you start searching for satellite broadband options, you might get excited reading about hefty data caps and download speeds over 10Mbps, all for well under $100/mo.

Wild Blue and Exede, two related consumer-focused satellite ISPs, have used spot beam from the start – making them totally unsuitable for mobile usage.

Don't be tempted to even try it if you want mobile flexibility.

If you want relatively affordable satellite internet you can take with you on the road, you need to skip over the newest technology and go old school.

There are two providers of "satellite broadband" that still maintain legacy systems that are potentially suitable for mobile users – HughesNet (www.hughesnet.com) and StarBand (www.starband.com)

All other consumer satellite options, including the modern offerings from HughesNet, exist primarily to serve fixed-location residential users in locations where traditional wired internet connections are not available.

123

Why Not Just Change Spots?

None of the consumer satellite systems support changing your assigned spot beam – it requires a certified installer on the phone with the network operations center to do so.

But these same satellites are being used to provide broadband service to commercial jets in flight and to mobile news trucks, and the robotic dishes in these systems do support automatically roaming between spot beams. Hopefully, in the future this technology will be more widely deployed, but right now the RV market is considered way too niche and budget conscious for anyone to invest in.

Satellite Broadband – HughesNet

The most popular satellite provider for RV satellite users, HughesNet, is moving to what they call Gen4 as rapidly as they can – which is a spot-beam service.

HughesNet is now only offering Gen4 equipment for new installations, meaning that if you want to set up on a nation-blanketing satellite, you absolutely need to track down a used modem.

This means finding an HN7000S modem – the currently offered HN9000 satellite modem will NOT work with the nationwide satellite signal, only with Gen4 spot beams.

Once you get a working HN7000S modem activated, you need to be careful to keep it. Every time you call HughesNet for support or activation, they'll try to send out an installer to upgrade you to the latest and greatest for free.

You don't want that – no matter what they offer!

HughesNet Options: MotoSAT Datastorm or Tripod?

Roof Mount

The MotoSAT company manufactured the Datastorm line of robotic roof-mount auto-aiming antennas (sold new for $5000+) to connect to the HughesNet network, and MotoSAT officially resold HughesNet service for a slight markup. This had been the gold standard of mobile connectivity for years, and the DatastormUsers.com online forum became (and still is) the best place online for discussing satellite internet.

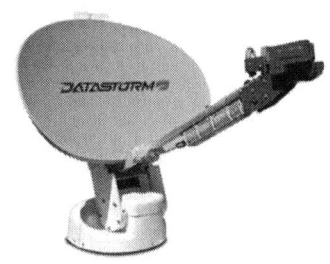

124

But MotoSAT went out of business in February 2013, and though some other satellite internet ISPs have stepped in to continue to service MotoSAT accounts (still via HughesNet behind the scenes), there are now very limited options for support and acquisition of equipment

Tripods

For a while there were multiple companies reselling basic home HughesNet systems, but instead of mounting the dish on the side of your RV they would provide training and sell you a surveyor's tripod and other gear to be able to set up and aim your dish yourself.

This sort of installation was vastly cheaper than a Datastorm roof robot, but was always in a bit of a gray area with HughesNet – making getting support a bit of a game.

The advantage of a tripod system was that it could be set up to avoid trees and other obstructions while you parked your RV in the shade (even a single small branch can block a data signal!). The downside is the lack of convenience – setting up and aiming a tripod could take 30 minutes to an hour, as opposed to pushing a button and waiting a few minutes.

A tripod system is not something you are likely to deploy to check email when you stop to fuel up or overnight at a rest stop.

HughesNet Systems & Resources

Equipment

If you are willing to fend for yourself, you can find old Datastorm and tripod systems literally being given away. Start with the Datastorm Users Group (www.datastormusers.com), various RVing forums, and check Craigslist and eBay. Nobody wants to ship big and bulky satellite gear, but if you are willing to go and help somebody take a dormant system off their roof, you can often get satellite for a song.

OregonRV.net (also runs motosatparts.com) buys up old equipment, and refurbishes and resells it to consumers. Their inventory will vary based on availability.

RFMogul.com is a group of former MotoSAT engineers who are working on some new Datastorm controller software and options.

Service

Activating service on a mobile setup directly with HughesNet can be an adventure all of its own. As of 2014, HughesNet is making it nearly impossible to activate new service on HN7000S modems, so expect a challenge.

They don't want to support mobile users, so you have to jump through hoops if you want to activate directly with them. You can try – but don't mention you're in an RV, insist you need legacy service, and don't let them insist on "upgrading" you to newer equipment.

A tip to try keeping your old used HN7000S modem is insisting that there is a line of sight issue with trees blocking the signal between your "house" and the newer satellite.

If you don't want to pull your hair out, you may be able to set up service through these resellers who, as of early-2015, all claim to still support legacy HughesNet HN7000S devices:

Montana Satellite – www.montanasatellite.com – Current cheapest HN7000S service plan is $109/mo for 1Mbps download speed and 350MB per day usage.

Mobil Satellite – www.mobilsat.com

Satellite Broadband – StarBand

StarBand is a much smaller satellite network provider than HughesNet, and though we've never heard of a robotic roof-mount antenna system that supports StarBand, some RVers have had good luck with tripod systems.

Unlike HughesNet, StarBand has not migrated to spot-beam technology – so you will not be pressured to upgrade. But StarBand also does not officially support RV deployments either.

StarBand reseller J2 Communications (www.j2communicationsltd.com), however, explicitly mentions RV support on their website, and MobileInternetSatellite.com offers StarBand tripod systems and training as well.

126

StarBand plans start at $49/mo for 512Kbps download speeds, up through $99/mo for 1.5Mbps. Weekly usage is capped, but usage is unlimited between midnight and 6 a.m.

A concern with StarBand is how rinky-dink the company appears. Would you trust your connectivity to a company that STILL claims that Windows Vista (released 2007) "compatibility testing is ongoing" and that Windows XP (released 2001) is the most recently supported operating system?!? (Never mind Windows 7 and now 8!!!)

Business-Grade Satellite Broadband

With MotoSAT gone, if you want full equipment, service, and support from a company that is still in business, there are several options – if you have the budget for it.

Ground Control (www.groundcontrol.com/) offers a "consumer-grade" Toughsat auto-aiming dish for $13,999 – and monthly iDirect service plans start at $399 with just 3GB of data included.

Another provider, Mobile Satellite Technologies (www.mobilsat.com), offers consumer data plans starting at $99/month for 600Kbps downloads and 60Kbps uploads, or for $260/mo you get 1500Kbps down, 200Kbps up, and support for VoIP voice calls.

BGAN (Broadband Global Area Network)

If you are more interested in basic connectivity rather than broadband speeds and bulk transfers, BGAN is a potentially interesting option.

Rather than a big bulky dish, a BGAN satellite antenna is about the size of a laptop computer – and it does not need to be carefully aimed and can usually be set up in a matter of minutes.

And BGAN service covers the entire planet, other than the poles, perfect for sailing nomads or expeditions who plan to roam continents. Traditional satellite service is usually limited to a single continent.

But...BGAN is slow and extremely expensive!

127

The most basic BGAN terminal like the Wideye Sabre 1 (www.groundcontrol.com/BGAN_Wideye_Sabre.htm) costs around $1,600 – and is capable of speeds of just 384Kbps down and 240Kbps up. Faster, fancier, and more rugged terminals go way up in cost.

384Kbps is not enough to even dream of streaming YouTube, but it is plenty for email checking and some basic old-school web surfing.

But it is the data costs that will kill you: At up to $6.99/MB, sending data via BGAN can quickly get expensive.

BGAN terminals can also double as satellite voice phones as well, making calls anywhere in the world for around $0.99/minute.

The key to using BGAN is extreme data-usage management – focusing on email and other text-centric communications rather than interactive graphical web surfing.

Low Earth Orbit – Globalstar & Iridium

Most satellite options rely on carefully aiming at satellites in fixed locations in the sky – parked in geostationary orbit over the equator 23,000+ miles away.

It takes a carefully aimed, traditional-looking dish to communicate over those extreme distances – or an expensive and slow BGAN terminal.

Satellites in lower orbits can be easier to talk to, but they are always in motion through the sky and will often be over the horizon and out of sight. But a constellation of satellites working together can provide constant coverage from much lower altitudes.

Globalstar

Globalstar has a constellation of 40+ satellites 880 miles up and has long been a provider of satellite telephone service. Now they are offering something new called Sat-Fi (www.globalstar.com/sat-fi/),  which they like to describe as a satellite-powered WiFi personal hotspot that supports unlimited data with plans starting at just $39/mo!

The Sat-Fi hotspot is $999 and supports eight users at once.

Before you rush to sign up for this perfect-sounding solution…

The catch is that the data speed is 9.6Kbps (vintage mid-1990s modem speeds), and the free data only covers basic email, SMS text messages, posting to FaceBook and Twitter (not reading), and very little else. The

$39/mo plan also includes 40 minutes of satellite phone voice minutes – which can also be used to make very slow general-use data calls.

For $149/mo – you actually get unlimited voice and data calls. The data speeds are so slow that you would never be able to interactively surf the web. But it you have extremely low bandwidth needs or are primarily voice-centric – this is actually a very affordable option.

Iridium

Very similar to Globalstar is the Iridium constellation, which provides global coverage with 66 swarming satellites orbiting 475 miles up.

Iridium has also launched a satellite WiFi hotspot in 2014, also priced at $999. Plans start at $49/mo, or you can get unlimited (but extremely slow 2.4Kbps) data for $134/mo.

At that speed, uploading a high-resolution image might take hours, and even a basic web page might take 5–10 minutes to display. This is not a device for surfing on!

The other big gotcha is that all data usage must go through Iridiums apps on your phone or tablet – you cannot get a laptop online or arbitrary apps.

DeLorme inReach

Another more consumer-friendly option that uses the Iridium network behind the scenes is the DeLorme inReach. The rugged handheld device costs $299 and provides continuous remote tracking and support for 160-character, two-way text messages that work essentially anywhere on the planet – with plans ranging between $12 and $99/mo.

If you are headed out on a remote expedition, this would be a great bit of technology to have with you.

SPOT Tracking

Very often when you are out in the boonies far from other communication options, your communication needs are actually very simple. You either want to let your loved ones know "I'm fine, and this is where to find me..." or you want to let everyone know "Send help – here is where I am!"

For these basic needs, a big expensive satellite system is overkill.

The basic SPOT Satellite Messenger (www.findmespot.com/en/index.php) device combines a GPS with a one-way satellite transmitter that works almost everywhere in the world (there are a few gaps in coverage, but not many).

Just press a button to "Check In" or another button to "Send Help."

The $170 SPOT Connect gives you some additional capabilities – when paired with your iOS or Android phone via Bluetooth, you can compose custom messages up to 41 (wow – novel length!) characters long that can be sent via SMS or posted to Twitter or Facebook.

SPOT service costs $99/year, and automatic tracking that logs your location to a shared Google map every 10 minutes costs $49/year.

The big downside of SPOT is that it is broadcast only – there is no way for you to receive a message. But it is affordable, and the safety and increased peace of mind can be very worth it for many travelers.

NOTE: Check the latest reviews on Amazon.com – there are a lot of mixed opinions about how well SPOT works.

Caps, FAPs, and Other Limitations

When you are using most satellite internet services, keep in mind that you are relaying everything via a satellite parked hundreds or thousands of miles above you.

That is a long way away, and the speed of light starts to impact everything in ways that feel completely foreign to those of us used to terrestrial connections.

Even with a fast connection, there will be a brief pause with every click as you wait for your request to make the trip from your dish, to the satellite in orbit, back down to the web server you are talking to, and then a full round trip back to you.

This is called latency, and it is especially noticeable if you try to have a video or audio conversation with someone. It actually helps to say "over" to avoid talking over each other.

Consumer satellite plans (like HughesNet) also commonly have data caps – however, they're quite different than cellular data caps. With satellite, you have a relatively small daily limit that refills throughout the day. If you exceed it, you're FAPed (Fair Access Policy) and put on lower throttled speeds for 24 hours.

But since your data bucket slowly refills throughout the day, you can avoid getting FAPed by stepping away from your computer. If you go take a hike or go out to dinner, by the time you come back you will have more bandwidth available.

Some HughesNet plans now include a token or two that you are issued each month, which you can redeem to reset your FAP counters. You may have the option of buying extra tokens, which is very flexible if you have a couple of days in which you need a lot of bandwidth to complete a project.

Another advantage is that on many satellite plans, overnight is usually a period of time with unlimited usage – ideal for doing larger file transfers and system updates. If only cellular plans offered such an incredibly useful feature!

Even Satellites Have Coverage Maps

The great advantage of satellite Internet service is that you can connect anywhere you have an unobstructed view of the southern sky. But there are actually still potential coverage map issues when it comes to satellite.

HughesNet, for example, offers service on a dozen different satellites, each with a different broadcast footprint and varying signal strengths across the nation. Not every satellite actually serves every corner of the country, particularly if you want to travel to Alaska, throughout Canada, and down into Mexico.

After comparing all the maps posted at DatastormUsers.com, we selected Galaxy 28 (www.datastormusers.com/glossaryterm.cfm?phrase=G28) for our former HughesNet account – which could give us connectivity while traveling into both southern Canada and partway down into Baja, Mexico.

Changing satellites on HughesNet is possible, but cumbersome. It is better to do some research in advance and try to get assigned to a satellite with coverage in the places you are most likely to go.

New Developments in Global Satellite Broadband

Things may be relatively bleak in the satellite world right now, but they are starting to get interesting again.

As of early 2015 - Elon Musk's SpaceX has launched a new initiative to develop a "global communications system that be would be larger than anything that has been talked about to date" - with Google as a key investor funding the project.

Not to be outdone, billionaire Richard Branson and Virgin Galactic are backing another new satellite internet startup, OneWeb, that is "creating the world's largest ever satellite constellation."

131

Both of these new initiatives are aiming to put hundreds of very advanced satellites into a low earth orbit swarm, low enough to avoid the latency issues of higher altitude satellites.

These two companies have the potential to revolutionize satellite broadband, but not overnight. The earliest either of these networks will be live is 2018, and even the most basic details will likely not be publicly revealed anytime soon.

But we can look to the skies and dream!

Mobile Satellite: Concluding Thoughts

A decade ago, satellite made sense. HughesNet could run circles around any cellular plan in both speed and coverage. Now the tables are turned, and cellular coverage is extensive and fast.

Satellite just can't keep up.

New technology may come along and bring a new generation of mobile broadband into orbit, but for now – unless you're planning extensive time away from cellular towers and WiFi options – satellite internet for RVs probably just doesn't make sense.

Alternative Connectivity Options

Cellular, WiFi, and satellite are not the only ways to get online while enjoying life on the road. If you are willing to get a little creative or experimental, there are a few other rarely considered alternatives too.

Particularly if you are willing to get a little flexible with what mobile and internet means, you may have more options than you ever imagined.

Amateur Ham Radio – Email & Internet

If you're a licensed ham radio operator with the right equipment, you can get access to Winlink (www.winlink.org).

This service allows you to send noncommercial emails over the amateur radio frequencies – which can be useful if you're out in the middle of nowhere with no other options.

The important thing about amateur radio is to remember that it is strictly for amateur use – it is against the law to use the amateur radio bands for any commercial purpose (including checking work email) or to transmit any encrypted data.

It also means that for the most part you need to be comfortable figuring things out on your own – commercial companies aren't going to hold your hand and coach you through the process of getting online. But if you are interested in pursuing the ham hobby, there is an incredibly vibrant and experienced online community very willing to mentor newbies.

If tinkering with radios turns you on, this can be a great free way to get online. Even with the inherent limitations, as a ham you will be able to manage some very basic data connectivity on the go, wherever you are.

133

And if you have a big enough ammeter radio rig and the appropriate antennas, you can usually get connected from just about anywhere on Earth.

An additional ham resource we're aware of is Broadband-Hamnet (hsmm-mesh.org/) – an experimental, high-speed, wireless broadband network that uses WiFi networking gear with custom firmware to build a self-configuring mesh network. If you are in an area with other users, you can be part of creating a widespread network that stretches much further than a single user alone could ever communicate.

If you want to get involved with amateur radio, the place to start is the ARRL (www.arrl.org) – the national association for amateur radio.

Cable / DSL Installation

Being mobile doesn't necessarily mean having to use only mobile internet!

If you're planning to be in one spot for a while, sometimes hooking directly up to fixed, wired cable or DSL is a possibility.

RV parks and even mobile home parks with RV spaces that cater to long-term residents and permanent dwellers sometimes already have cable pulled to each site, and all it takes is contacting the local cable or DSL provider to get service switched on and have them bring you the necessary modem.

The park may not advertise this as an amenity, so you probably need to ask. Start by looking for parks that offer cable TV or seasonal rates.

Depending on the provider, the costs to get started are very reasonable.

Since there are often not any long-term contracts required to get service, you can cancel after just a few weeks or month or two without penalties.

And you can usually rent the equipment for a few dollars a month, instead of buying it. But if you find yourself signing up for cable internet often, many providers utilize the same modem standard – so it may be worthwhile buying a cable modem and keeping it onboard for quicker activation.

The advantage of going with cable or DSL is gaining access to fast and essentially unlimited internet. After rationing out bandwidth on the road for

134

months on end, spending a few weeks drinking from the firehose can feel incredibly decadent and absolutely awesome!

Temporarily embracing a wired lifestyle can sometimes be very worth it for us bandwidth junkies – perfect for hyper-focusing on a work project before heading back out on the road!

Caution: When you rip yourself away from unlimited fast bandwidth again, you may whimper.

WISP Access (Wireless Internet Service Providers)

Particularly in the mountainous West, a WISP may be an option for temporary fixed-location service, even in places beyond where cable and DSL providers reach.

A WISP is a wireless internet service provider – and these companies have sprung up in many communities to fill the demand for faster-than-dial-up home internet service. The WISP providers set up transmitters on local high points and then install compatible broadband receivers on the roofs of local customers.

Since the WISP doesn't need to run new wires, if you have a view to the right mountain or hillside, you might be able to get a local WISP to offer you fast unlimited service – even if you are in a remote boondocking spot miles from the nearest cable run or phone line. It all depends on line of sight.

To find out if WISP might be an option in your area, check local advertising periodicals, signs in grocery stores or laundromats, the Yellow Pages, Google, or talk to local computer repair professionals for leads.

Co-Working Spaces

Co-working spaces are office setups for independent workers to base themselves out of. They offer amenities like a desk, connectivity and meeting rooms. Sometimes they can be far more productive than trying to get work done from a cafe, as everyone around you is also working.

They're located all over the country, particularly in larger cities. Some of them offer short term rental options including hour and day passes, or even monthly space. These can be an ideal option for a mobile worker to setup a work base camp for a bit.

Start with resources including CoWorking (wiki.coworking.org), ShareDesk (www.sharedesk.net) and Wherever Worker (www.whereverworker.com) to located co-working spaces available.

Concierge Services and Personal Assistants

Sometimes the best way to get stuff done online is to let someone else do it for you.

Think about how long it might take you to research a topic or make a reservation with a web connection that bounces up and down more often than a yo-yo, and a sporadic voice signal that limits conversations to "Can you hear me now?" and "Wait, what was that?" and "#%$*@! stupid phone!"

What if instead you had someone with a rock-solid internet connection willing to do your bidding on demand, and who will get back to you later with the results or needed information?

It may seems counterintuitive at first, but consider these options...

- **Tap into Online Community** – Rather than fighting with a slow connection for hours researching options, try instead to get online just enough to post a request for advice and information to your Facebook friends, your blog, or to to a relevant forum you participate in.

  Go for a hike, return a few hours later, and you often may find yourself with a wealth of information waiting.

  This is a great way to get routing recommendations or general technical advice. And it sure beats hours of cursing at web pages slowly loading! Just be sure to carefully weed out the advice you don't need – you will find everyone has an opinion, and not all are appropriate for your situation. And, of course, return the favor for others when you have a good internet connection and someone needs a little information.

- **Friends/Family** – If you have specific requests and tasks that need to get handled and your lack of good connectivity is making you pull your hair out, don't be afraid to get a message out to a friend or family member asking for help, particularly if you are dealing with an urgent or emergency situation.

  And don't forget to thank them. Maybe a photo from your travels or a gift?

- **Credit Card Concierge** – There is a chance that you might already have a personal assistant on call and not even know it! Some credit cards (such as Visa's Signature cards and many AMEX cards) offer a concierge perk, and it can actually be surprisingly useful.

136

We once found ourselves camped out in the boonies with minimal data and voice service, and urgently needing a dentist. Rather than struggle with limited signal to find a dentist who could see us on short notice, we got a call through to Visa and tasked them with finding a nearby dentist willing to see an urgent new patient that very morning. Visa called around and even made us three tentative appointments.

- **Online Virtual Assistant Services** – If you have tasks that are more complicated than a free concierge can handle or you want a concierge who "knows you," there are a lot of online personal assistant and concierge services that you can subscribe to. Check out Ask Sunday (www.asksunday.com/), which assigns you a "Dedicated Assistant" with monthly plans starting at $119/mo, or an on-demand service for just $29/month. Fancy Hands (www.fancyhands.com) handles smaller requests in packages as low as $25/month.

- **Dedicated Personal Assistant** – And finally, some nomads have found it worthwhile to actually have someone on staff. If you are running a business and need to keep up appearances of always being available, having someone who answers the phone, triages replies to emails, and handles making reservations for you can buy a lot of flexibility and freedom.

Having a fast and reliable assistant can ultimately minimize your needs for staying close to fast and reliable internet.

Routers – Bringing It All Together

A router is the juggler in the middle of your network – tying together all your devices and all your connectivity options, acting as the funnel that all your internet traffic passes through.

Routers can be so incredibly basic that you may not even realize you have one, or so advanced that you may feel like you need an IT person sleeping on your couch to manage the options.

It all depends on what you want to do and how much you want the router to juggle.

Do you have multiple devices that you want to be able to simultaneously share a single internet connection?

Do you need a way to more actively meter and manage your network usage?

Would you like to be able to open your network up to invited guests, without compromising your security?

Ultimately you probably will want a router in your tech arsenal – and you might be surprised to discover that you already have one.

Do You Really Need a Router?

The most basic function of a router is taking an upstream network connection and sharing it with multiple downstream devices over either wireless WiFi or wired ethernet.

If you never intend to share a single connection across multiple devices, you do not need a router. A simple USB modem plugged directly into a laptop or a direct connection on your phone or tablet can actually save you a ton of headaches.

And if you only have a few devices that you want to share a connection, you can probably just use the routing capabilities built into the personal hotspot feature on your smartphone.

If, on the other hand, you have an entire collections of devices that you'd like to get online, as well as enable them to talk to each other, you almost certainly would benefit from having a more-powerful router sitting at the heart of your network.

In our case, even our lightbulbs and bathroom scale are WiFi enabled, and a router is an absolute necessity!

Router Selection Tips

Some nomads use traditional home wireless routers, such as the Apple AirPort or any of the vast range of routers sold by LinkSys, D-Link, Netgear, and many others.

These home routers, however, do not have any built-in support for cellular data connections, making them substantially limited for mobile use.

Some general consumer routers and WiFi range extenders support WiFi-as-WAN functionality and a whole raft of other features, but they are all designed for extending WiFi networks where you control both ends of the connection – a situation that is hardly ever true for an RV traveler. You will save yourself some headaches by going with something designed with the mobile user in mind.

Here are the different types of mobile-optimized routers you will come across.

Phone/Tablet Personal Hotspot

A lot of smartphones and tablets have built-in personal hotspot features that let the device actually act as a wireless router, sharing its network connection and creating a WiFi hotspot.

Every smartphone platform refers to this feature differently. Apple calls it "Personal Hotspot," Androids tend to call it "Tethering" or "Portable Hotspot," Windows phones call it "Internet Sharing," and BlackBerries call it "Mobile Hotspot Connections."

Once you flip the switch on this feature, your other devices will see a wireless network that they can then join.

140

This type of hotspot is usually limited to serving just a few connected devices at a time, usage burns through your host device's battery rapidly, and if you are idle too long the hotspot may go to sleep and vanish until you restart it.

This is the simplest type of router you will run across.

Mobile Hotspots (aka MiFi or Jetpack)

The next step up in router complexity is a small, battery-powered cellular router, usually around the size of a deck of cards.

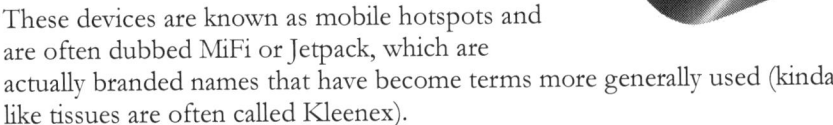

These devices are known as mobile hotspots and are often dubbed MiFi or Jetpack, which are actually branded names that have become terms more generally used (kinda like tissues are often called Kleenex).

These gadgets are essentially just small, battery-powered routers with a built-in cellular connection. When turned on, the MiFi connects to your network provider and then creates a local WiFi hotspot for 5 to 10 nearby devices.

The advantage of a MiFi is how simple they are to use. They do not, however, typically support advanced features or support multiple upstream connection options.

Cellular & WiFi-as-WAN Routers

Though it ends up being a more complex setup than a MiFi, often there are advantages to using a specialized router that can bring together multiple upstream connection options.

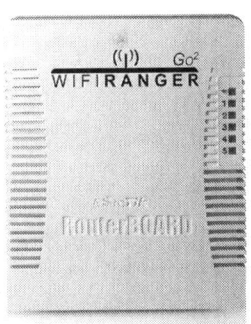

The key feature to look for is direct support for controlling cellular modems and for a feature known as WiFi as WAN.

WAN stands for Wide Area Network, and on a router this is the term for the upstream connection. WiFi as WAN is the capability of a WiFi router to simultaneously create a local private WiFi network while also connecting upstream to a public WiFi network.

This scenario is common for RVers connecting to campground WiFi.

A mobile router may also be able to juggle other upstream data sources, such as a cable modem or satellite connection – and you should look for failover features that define how these connections are prioritized and automatically switched between.

141

One common configuration is to set up your router to automatically connect via WiFi as WAN to any open WiFi network, and failing that, to automatically switch to a cellular connection. This way the free and unlimited connection becomes the priority.

The nicest thing about using a router like this in the heart of your networks is that it keeps things simple on all your client devices. You just need to configure all your devices to point to your router's hotspot, and when you change locations you do not need to reconfigure any of your devices or reenter any new passwords.

Here area some of the current router options suitable for RV use:

WiFiRanger: WiFiRanger (www.wifiranger.com) is a small company focused on providing WiFi/cellular routers for the RV market. Because of this focus, WiFiRanger has invested a lot of effort into keeping things as simple as possible for nontechnical users, while attempting to solve pesky problems like getting the router past campground login pages.

WiFiRanger has various models designed to work well together. The Go^2 ($239) is a relatively inexpensive mid-power indoor router, and the Elite ($359) and Sky ($399) are higher power devices designed for permanent outdoor installation. When connected together, they automatically work together via one interface.

For most RV users with serious WiFi needs, the WiFiRanger is a great all-around choice.

Pepwave: Pepwave (peplink.com) is primarily focused on the enterprise market and produces a whole range of advanced and extremely capable routers. The most consumer-friendly and affordable product in the Pepwave lineup is the Surf SOHO 3G/4G Router, which retails for under $150. You can get them at 3GStore.com, amazon.com and other retailers.

The SOHO offers a solid feature set and really excels at tracking data usage.

Cradlepoint: Cradlepoint (www.cradlepoint.com) has been backing away from the general RV and consumer-friendly mobile-router business to focus on higher end markets, and Cradlepoint's most appealing general-purpose device, the MBR95, is orphaned and will no longer be receiving software updates.

142

But like Pepwave, Cradlepoint has a whole range of higher end mobile-office gear that remain top-notch choices for those with demanding needs. The MBR1200 ($250) and MBR1400 ($380) are particularly appealing high-end options. But be aware that now all Cradlepoint models require an annual support contract fee to be paid to receive ongoing firmware updates and support.

TechnoRV: TechnoRV(www.technorv.com) sells a WiFi USB Repeater ($99) that allows their various USB WiFi adaptors ($59–$139) to be shared over WiFi instead of with just a single computer over USB. This is only suitable for setting up a WiFi-as-WAN connection – this router does NOT support using cellular USB modems or failing automatically between connection options.

This is a cheaper but more limited option than WiFiRanger, but some RVers have had good WiFi results on a tighter budget with this setup.

Advanced Configurations

Depending on your connectivity needs and tolerance for or enjoyment of network hacking and configuration juggling, you can build up an even more complex router configuration.

One good place to start is by checking out DD-WRT (www.dd-wrt.com), an open source alternative router firmware project that lets you tweak, configure, optimize, and automate router functionalities.

If you have a supported router model and are willing to void the warranty by replacing the factory programming with something more complex, serious geeks can have a lot of fun here. Some routers can now be purchased with this firmware pre-installed.

But even serious geeks often prefer to have something more simplified, tested, and supported.

You can get many of the same advanced capabilities and more by going with a high-end, commercial-grade router from Cradlepoint, Pepwave, or others.

One feature common in higher end routers is load balancing – the capability to use multiple upstream connections simultaneously to spread out connections across all the available devices, balancing usage. WiFiRanger has added load balancing to the new Phantom 7.0 firmware for WiFiRanger routers, released as a free upgrade February 2015.

Connection bonding takes load balancing even further – creating a VPN connection that combines multiple upstream paths into a single faster virtual connection. For connection bonding to function, you need to have a server reversing the process at the other end – usually another router from the same manufacturer.

A well-configured router with load balancing enabled can actually take into account network speeds, data caps, and costs – automatically attempting to optimize your usage across multiple carriers. Pepwave even has routers with integrated cellular modems featuring multiple-SIM slots, allowing the router to automatically switch to a secondary account when all the data on the primary account is used up.

Higher end routers also often support integrated VPN functionality – allowing your entire mobile network to virtually appear local to a remote network. If you are working remotely for a larger corporation, the IT staff may insist on this sort of configuration.

WiFi Congestion Issues

Most WiFi devices operate on an unlicensed 2.4GHz frequency band that provides for only THREE fully distinct radio channels (en.wikipedia.org/wiki/List_of_WLAN_channels)

> TIP: If you are manually configuring a WiFi device, channels 1, 6, and 11 are the ones that do not overlap and interfere with each other.

Bluetooth devices and most cordless phones (remember those?) also operate in this same 2.4GHz frequency band – further adding to the congestion. In an urban area or even a crowded campground, those three channels can get awfully overloaded with signals, all interfering with each other.

But it is even worse than that: Microwave ovens also emit radiation in the same 2.4GHz band, meaning that if your neighbor is making popcorn, it can potentially grind WiFi speeds to a halt for everyone nearby if the microwave isn't well shielded.

Despite all the interference, it is amazing how well 2.4GHz WiFi works.

But with so much traffic and more and more WiFi devices in use every day, using WiFi on 2.4GHz is like trying to have a conversation in a loud bar while a heavy metal band is playing and a jackhammer is in use on the street outside.

5Ghz WiFi – An Uncrowded Expressway!

There are also WiFi channels located in the uncongested 5GHz frequency band, where 23+ nonoverlapping channels are sitting usually vacant. It is the express lane compared to the 2.4GHz gridlock – a quiet library compared to a raucous bar. But to take advantage of 5GHz requires different antennas and equipment specifically designed to broadcast on these channels, and even now many new WiFi devices skimp on 5GHz support.

It has been a classic chicken-and-egg problem – manufactures need to include 2.4GHz support for backwards compatibility for communicating with existing devices, and since there are so few 5GHz devices, manufacturers are tempted to keep costs lower on new devices, leaving 5GHz support out.

WiFi is defined by the IEEE 802.11 (en.wikipedia.org/wiki/IEEE_802.11) set of standards – and the specifications for any WiFi-compatible device should indicate which variants (indicated by letters appended to 802.11, such as 802.11b/g/n) of the standard are supported. When two WiFi devices try to connect with each other, theoretically they should negotiate the fastest and most recent connection standard that they are both compatible with.

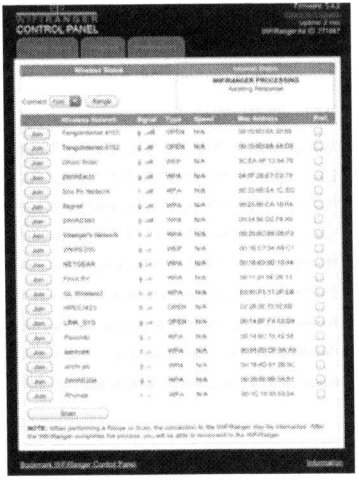

The 802.11b and 802.11g WiFi standards run at 2.4GHz only.

802.11a is 5GHz only.

The now increasingly common 802.11n standard supports both 2.4GHz and 5GHz frequency bands, but 5GHz support is optional, and many 802.11n routers only support 2.4GHz. And some that support 5GHz only allow one frequency band to be used at a time – meaning that enabling 5GHz means locking out all your 2.4GHz equipment!

The brand new ultra-fast 802.11ac is designed to really take advantage of 5GHz, but most 802.11ac gear remains backward compatible with 802.11n and 2.4GHz devices too.

Simultaneous dual-band 802.11n home routers are becoming increasingly common. Unfortunately, we have yet to find a consumer-priced, mobile-optimized router that also supports cellular data cards, or any mobile MiFi hotspot, with simultaneous dual-band support.

145

Since our WiFiRanger router is 2.4GHz only, if we want a fast, uncongested, reliable network between our gadgets that means adding a second 5GHz-compatible WiFi router – or reverting to a wired network where possible. We chose to go with a wired local network. Particularly for streaming HD video, 2.4GHz WiFi often just doesn't cut it.

TIP: If you are buying a WiFi router, look for 802.11n (or 802.11ac), explicit mention of 5GHz, and especially "simultaneous dual band" support, which allows both frequency bands to be used at once.

It may cost more, but it is worth it if you want to actually take advantage of the wireless speed that you are paying for. And if you are buying a laptop, make sure that 5GHz WiFi or 802.11a is supported. For years now, all Apple laptops and iPads have had dual-band support, but the iPhone only added support for 5GHz as of the iPhone 5.

Some of the latest Android flagships have already made the jump to support 802.11ac already too!

Wired vs. Wireless Speeds

Plugging your fancy laptop into a wired network jack may seem like a step backwards into the primitive pre-wireless age, but especially if you have multiple computers talking to each other (and not just upstream to the internet) it can actually end up making a lot of sense.

Consider – even if you have a 5GHz 802.11n wireless network, the maximum theoretical speed is usually at best 300Mbps – and in less-than-ideal conditions, the speeds will be much, much lower.

Meanwhile, a wired gigabit ethernet network can run at a full 1,000Mbps speed no matter what, in both upstream and downstream directions simultaneously.

Since we have our backup and media drives attached to a Mac Mini acting as a server, having a wired network means that these drives are reachable from the other computer nearly as fast as a local hard drive would be.

If you do build a wired ethernet network, make sure to seek out hardware that supports gigabit speeds: "Fast ethernet" can barely outrun 802.11n WiFi. You will also need a gigabit switch, which can be had cheaply. Don't rely on the ethernet ports in your router, though – a lot of routers seem not to have gigabit switching built in.

Apple laptops have always supported gigabit ethernet speeds, but PCs are more hit-or-miss. Check your specs to be sure.

Managing Mobile Bandwidth

A mobile internet connection benefits from careful management.

Unlike typical home connections, which are uncapped and unthrottled and rarely changing, mobile internet connections vary in quality and speed as you travel and are always at risk of running into potentially expensive overage charges or frustrating speed throttles.

A little effort spent keeping tabs on your connection speeds and usage can go a long way towards maximizing your online experience.

Network Speed & Quality Testing

If you are going to invest any effort in optimizing your mobile data connection, you need to have ways to measure the impact your changes are having.

Focusing on more bars is not enough. Using boosters and repeaters to improve the signal strength only tells you half the story.

To really evaluate a mobile network connection, you need to keep a close eye on your actual speed and latency testing results. Here is how you do it.

Speed Testing Services

There are numerous speed testing services and apps – these are the ones we regularly use:

- Ookla Speedtest (www.speedtest.net)
- Ookla Speedtest App (www.speedtest.net/mobile/) – For iPhone, iPad, Android, and Windows phones.

147

- DSLReports Speed Test (www.dslreports.com/speedtest)

- Speed Of Me (www.SpeedOf.Me)

If you ever get results that seem odd, try another service.

Understanding Speed-Test Results

You will get three results from most speed tests:

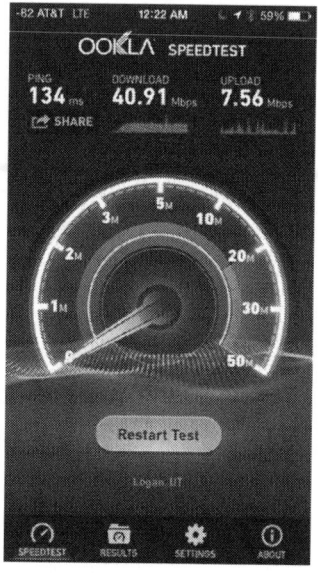

•**Latency (aka Ping):** This is the time in milliseconds it takes for a request from your computer to reach the speed-test server and to return, like the ping of a ship's sonar. The higher the number, the slower the speed.

Latencies under 100ms are good, under 50ms are great, and over 500ms begin to feel painful. This measurement is particularly important for online gaming – but any interactive task can begin to suffer with higher latencies.

•**Download Speed:** Reported in either kilobits per second (Kbps) or megabits (equivalent to 1000 kilobits) per second (Mbps). This is a measurement of the maximum speed that data is able to flow to you from the speed-testing server. Speeds over 3Mbps give a good surfing experience, and over 10Mbps will feel awesome. Speeds under 1Mbps start to make the modern internet feel slow, and speeds under 500Kbps are painful.

Download speeds have a particularly huge impact on streaming audio and video – if the speeds aren't able to keep up, you will experience stuttering, pauses, and long buffering delays. If your speeds are below 500Kbps, don't even waste any time trying to use video sites like YouTube.

•**Upload Speed:** The opposite of download speed – how fast is data able to get from you to the speed-test server. Upload speeds are almost always substantially lower than download speeds, and for many typical internet tasks upload speeds don't have a huge impact. But upload speeds are critical for video chatting and, of course,

148

uploading large files like photos, videos, or cloud-synced backups. Speeds over 500Kbps are the bare minimum for video chat; speeds over 1.5Mbps can deliver smoother results.

Speed Test Tips & Tricks

Speed tests can vary a lot from moment to moment – and a lot of that variability may not have anything to do with your personal network connection. To get a sense for the actual health of your connection, running several speed tests over the course of 10 or 15 minutes can help you get a better sense of what average speeds you are actually achieving.

Most speed-testing sites and apps have a way to change the server – letting you select a different server to communicate with and test against. Trying different servers can help you rule out whether strange results are isolated or not.

If you are comparing usage between two devices, make sure that your speed tests are using the same server, though! It is easy to have an overloaded server make one connection test slower than another, when in fact it might actually be faster.

And finally – keep an eye on your data usage. Speed tests work by sending large chunks of data back and forth to the server, timing how long it takes. Excessive speed testing can burn through your monthly data bucket rapidly if you are not careful.

How Much Bandwidth Do You Need?

If you're not used to relying on capped or metered bandwidth, you may have no clue how much you actually use each month.

Getting a handle on your actual expected usage is critical when building your mobile arsenal.

To start with, you should do an assessment of your usage and monitor it for a while – even before you hit the road.

If your internet provider doesn't provide this number for you, it is recommended you install a bandwidth counter on your computers and/or router to record your monthly usage.

While you can go and get average figures for typical file types (hours of video, number of webpages, and average emails, etc.) – you'll probably be surprised by what your actual usage is versus your imagined usage.

> Us humans are really bad at guessing what we consume. This is kinda like calorie counting when preparing to switch diets. While we'd like to think we eat a well-balanced set of meals, all those little snacks in between add up too.

Track your actual regular usage for a reasonable amount of time (at least a full week or, better yet, a month) to get a baseline. Remember to factor in all of your devices that you'll be connecting on the road – the tablets, the smartphones, the music players, the game systems, the eReaders, and all the laptops and computers that will be in your household.

Also consider how you currently consume media content – like movies and TV shows. If you're doing that over cable TV now, how will that translate for you once you no longer have cable?

As of March 2014, researchers report that heavy video-streaming home users in the US can use over 300GB a month – on the road that will cost you BIG!!! An average "cord cutter" (someone streaming over the internet instead of subscribing to cable) uses over 200GB and watches around 100 hours of online video a month. An average non-cutter user consumes around 30GB with minimal streamed video usage.

How do these numbers compare to your own usage?

Now, compare your own personal baseline to what it would cost to buy that much mobile bandwidth.

You may be shocked by this number, and be realizing at this point that you're going to have to do some serious bandwidth trimming to make mobile surfing affordable.

Tracking Your Bandwidth Usage

To avoid surprise overages or accidental account suspensions, you need to keep on top of your data usage. Most internet providers give their customers a way to check usage directly with them – either via a customer login account online, an app, or sending a text message on smartphones. Check with your carrier for instructions on how to do this for each of your devices.

Set it up, and get in the routine of checking each of your internet sources throughout the month to make sure you're staying within your quotas.

Some devices also have built-in usage tracking right on the device. Many modern mobile hotspots make this really easy and display your usage and limits right on the control panel screen, or even via a companion smartphone app. The device retrieves the carrier's report of how much data has been used so far in the billing period for display.

If you don't have an easy way to tap into your carriers usage metering in real time, you'll need to track it on your own.

Tracking your own independent usage can also help you figure out if you have a rogue program sucking up bandwidth – such as a sync to the cloud or a large update downloading. And if there's an unexplained spike in usage on your carrier's reporting, you can go back to your logs and see if it's accurate or perhaps an erroneous report that you need to refute with the carrier.

There are multiple ways to independently track your usage.

On-Device Usage Tracking Tools

You can install a tracking tool on every internet-using computer or device to track usage and monitor speeds.

For Windows, check out:

- NetWorx (www.softperfect.com/products/networx/)
- BitMeter II (codebox.org.uk/pages/bitmeter2)
- NetSpeedMonitor (www.floriangilles.com/software/ netspeedmonitor/)

For Mac, check out:

- Activity Monitor (pre-installed with OSX) can provide data usage for the network since your last reboot.
- iStats Menus (bjango.com/mac/istatmenus/)
- SurplusMeter (www.skoobysoft.com/utilities/utilities.html)

The advantage of on-device tracking is that these programs can often track instantaneous usage per-app, helping you isolate the hogs.

It is difficult however for these programs to tell the difference between local area network traffic (such as between two computers on your network) and internet traffic that is using up your cellular data.

And the big limitation is that you need to total up your overall usage across all your devices manually, and these tools have no way of measuring off-device usage - such as a video game console's data consumption.

151

Router-Based Usage Tracking

To really know how much data all of your devices are using, you need to funnel all the usage through a single point where it can be monitored.

That's when a router comes in super handy.

Most routers have some sort of usage tracking capability - but how useful and well implemented these features actually are vary greatly.

The Pepwave Surf SOHO in particular has some impressive usage-tracking tools that can even help you isolate which devices on your network are the most hoggish, graphing usage by device over time.

WiFiRanger's new Phantom 7.0 firmware (pictured below), released in February 2015, adds in a bunch of truly robust bandwidth tracking, monitoring and load balancing features, bringing together some professional level features to a consumer device.

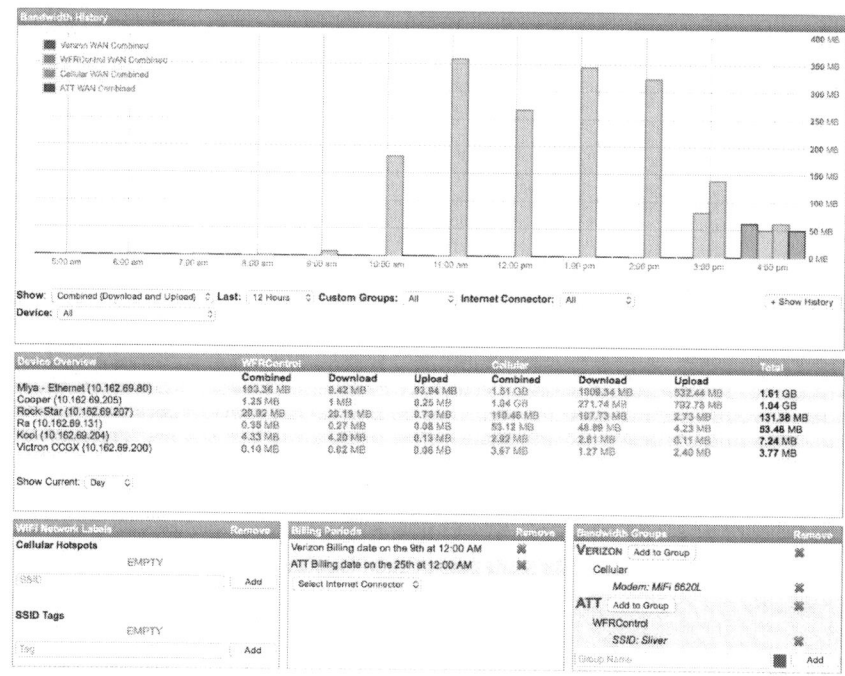

Android Usage Tracking

If you are using an Android device as your upstream connection - the Android OS has had a very powerful built in cellular data usage tracker since OS release 4.0 (Ice Cream Sandwich). This tool is found under the "Data Usage" section of the settings menu.

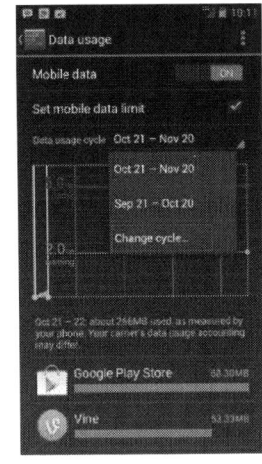

Not only can you see usage by app, you can set the dates of your monthly billing cycle, and even set warning and cutoff limits to prevent overage charges.

There are also dozens of other data tracking tools available for Android.

iOS Usage Tracking

Apple has more primitive built in data usage capabilities than Android, but you can still track overall and per-app data consumption under "Cellular Data" on the Settings menu. You can even block badly behaving apps from being able to use cellular data with just a tap.

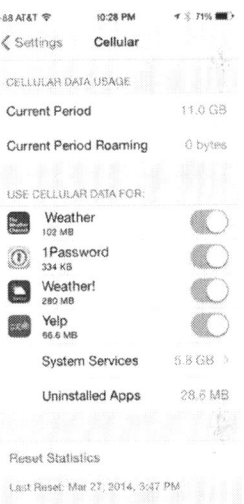

In case you are looking for it, the data used by the Personal Hotspot feature is hidden beneath the System Services selection.

The catch with the built in iOS data usage tracking is that you must manually reset the statistics periodically, making it hard to keep track of usage coinciding with your current monthly bill.

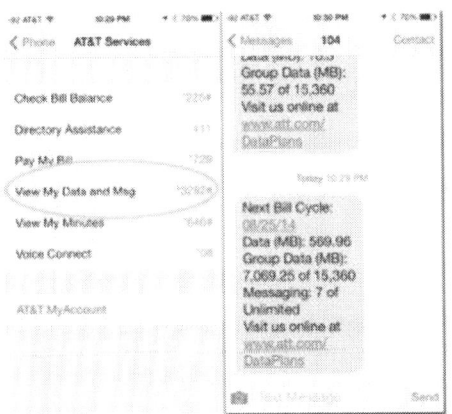

To see you current monthly usage, hidden at the bottom of the Phone menu on the Settings app is a place for your carrier to add special features. AT&T has used this to provide a button that will request AT&T to send you a text message with an update on usage in your current billing cycle, showing both your personal usage as well as the total used and remaining data on a shared plan.

153

Minimizing Data Usage

When you're on mobile data caps, you have to learn some tricks to minimize your data usage. In particular, LTE data gets burned quicker than most folks realize.

Here are some ideas to minimize that amount of data you burn through:

- Unless you need the speed, you may be able to force your device to connect via 3G/CDMA instead. Here's instructions on how to do so. (blog.wirelessworld.com/how-to-turn-4g-back-to-3g/) For a lot of stuff, 3G is fine, and the slower speeds will help minimize your data usage. On the other hand, 3G networks are starting to get pushed to the back burner and performance may suffer more than you can bear, and a lot of devices no longer have the option to force your connection to 3G when LTE is present.

- Be very careful when you load a page with video on it. If it autoplays, it's very likely caching faster than you can watch it. Meaning if you click away partway through the video, you've already spent the data – regardless if you actually saw the whole thing. Even if you click "Pause" or "Stop," it still caches in the background.

- Another video gotcha: Sites will often automatically play back HD video if you are on a fast connection, rapidly burning through data!

- Make sure your bookmarks to sites you visit frequently go directly to the actual page you start your experience with. You may have a bookmark to your bank's website, not to where you log in. Or a link to a forum you participate in – but the link goes to the front page, not the listing of recent postings you like to start with. No sense wasting bandwidth just to load a welcome page!

- Make sure you have auto-downloads of system and software updates turned OFF. Save those for when you have unlimited internet! A big OS update might be gigabytes in size, in one download using up all your data for the month. Wait and download these later.

- Watch out for iOS updates! In the past iOS devices have been known to automatically download OS updates when on a WiFi connection, with no way to manually opt-out other than staying disconnected from WiFi. If your WiFi is provided by a cellular hotspot, this unwelcome update could push you into overage territory when a major new release comes out.

- Pause auto-syncing to your cloud-based backup services – like DropBox. (CrashPlan apparently has features to help with this.)

- If you are subscribed to podcasts, TV series, or any other periodical content through programs like iTunes, be extra vigilant that you are not auto-downloading new episodes in the background.

- Close programs that sync or upload to other services that you're not actively using, such as iPhoto to Flickr, iTunes downloading app updates and podcasts, Adobe Creative Cloud product updates, programs that you have saving to a cloud service, and more. Even if you think you have the settings tuned just right, things can easily get messed up – sometimes causing repeated attempts at partial syncing and uploading.

- If you have multiple computers and devices, make sure that they aren't duplicating downloads. If you are subscribed to a podcast that you want to download new episodes of, make sure that you are doing so on only one of your devices. No sense having three copies of something!

- Run an adblocker in your browser to avoid loading unneeded graphics and promotional video ads. Disabling Flash support can make a huge difference.

- A lot of apps, unless they are specifically designed for mobile usage, are built with the assumption that data is unlimited. Even when they appear to be idle, some apps are actually burning through data in the background - downloading updates, content that you may never even view, or even updating advertising for quicker display later. Be careful of what you leave running in the background!

- And if you're not using it, turn your internet devices OFF or disable internet on your computer to prevent background tasks from silently eating away at your data allotment.

- Game consoles, Blu-ray players, and even some smart TVs can use up to a ton of data silently in the middle of the night downloading updates. Do not leave these devices plugged in to mobile internet unless you are keeping a close eye on them.

- If you are an advanced user, you can set up your own proxy server that you route all your connection requests through. The proxy then can compress the data before it gets transmitted, potentially even recompressing graphics to be lower resolution or throwing away data that the proxy feels is unimportant.

  Some cellular and especially satellite internet connections actually do this for you automatically in the background – you may sometimes notice that pictures look slightly worse when on cellular.

- Note, though, that a proxy server (whether your own or via your cell provider) can only compress and modify unencrypted pages. If you

155

are visiting an encrypted site, no one in the middle can modify what is transmitted (for good or for ill).

- Use Chrome's Bandwidth Management: To start saving data using the Chrome for Mobile browser, visit Settings > Bandwidth Management > Reduce Data Usage. Then simply turn the toggle to "On." From this menu, you'll also be able to track how much bandwidth you save each month as you browse on Chrome. A lot of cellular carriers do the same thing behind the scenes too.

No matter how careful you are, you will still inevitably stumble into accidentally using way more data than you realized or thought possible.

It is tempting to blame your internet provider and to disclaim all personal responsibility. We hear it all the time - "there's no way I could have used that much data!" Heck, we've said it ourselves when notified of high usage.

But in reality, screwed up accounting behind the scenes is only occasionally to blame. Very often some forensic digging finds the real guilty party. We have seen OS updates and iTunes ignore "auto download" settings, upgrades to Google's Picassa photo manager silently opt-in to "backup all photos to the cloud," iPhone/iPad OS updates auto download when connected to WiFi (even over Jetpacks) and even an Xbox One turn itself on in the middle of the night to download a huge 4GB update with no disclosure or warning given.

If you ever accidentally blow through your data caps on your primary account, it is extremely useful to have a secondary way online to help you limp through to the end of the month when your caps reset.

Scoring "Free" Data

A potential big shift in how mobile networks work and are paid for is called "zero rating." Zero-rated content does not count against your data allotment for the month. The rate charged to transmit the data is thus "zero" for the end users.

This is like an 800 number used to be on a voice phone network, but it puts any sites that aren't zero rated at a huge competitive disadvantage. Would you still use Google if it cost you, but Bing was exclusively partnered with your mobile internet provider and was thus free? Should a company like Amazon be able to cut deals that give Prime subscribers free unlimited streaming on one network, while another network might get free Netflix?

AT&T has been making moves to allow for "sponsored data" on their wireless network – usage that is zero rated and does not count against your monthly data caps.

A company called Syntonic Wireless (www.syntonicwireless.com) is rolling out an iOS and Android app that will act as a portal to sponsored content on AT&T – allowing for a range of sites to be visited (only through the app) without it counting against your monthly cap.

Sites that have signed up with Syntonic include Airbnb, Amazon, eBay, ESPN, Etsy, Expedia, Facebook, MLB.com, Open Table, Rolling Stone, the *New York Times*, and Yelp – with more on the way.

On one hand, this is great for users fighting to stay under the monthly caps – but on the other, this is a genuine threat to the idea of an open internet – with nonparticipating sites left at a competitive disadvantage, and AT&T left with less incentive to bring out lower prices for nonsponsored data.

It is also unclear how much information the sponsoring sites get handed about you in return for paying for your access.

Another similar new service is called Kickbit (www.kickbit.com) – which is an iOS and Android app that lets you earn data by "engaging with brands" and "completing exciting offers." So far they have had offers like "714MB earned for ordering flowers via FTD" and "20MB earned for tweeting about chocolate."

The latest Kickbit offer is amusingly ironic – encouraging you to "sign up for a 7-day trial of Hulu Plus, and earn 200MB." Of course, that will be gone in less than an hour if you actually try to watch Hulu Plus streaming video on your mobile device!

AT&T, Verizon, Sprint, and T-Mobile are all supported by Kickbit. This trend towards sponsored data seems to be gaining steam – and there will probably be more similar options soon.

Entertainment on the Go

Many home-based entertainment options have evolved to consume a lot of internet data – streaming video, online gaming, or even just general web surfing. And it seems that content creators and providers are always increasing the quality of their offerings, which means they eat up even more bandwidth.

If you have fast and unlimited bandwidth, like cable internet, then this is no big deal.

But for those of us managing mobile bandwidth with capped data and variable speeds, we quickly start running into problems if we're not willing to adjust our expectations and viewing habits.

With some forms of mobile data being as fast, or faster, than home-based cable services, you can use your data up quicker than you might think by watching just a couple movies online.

Watching it on public WiFi hotspots – such as at campgrounds? That's going to be highly variable. Keep in mind, most WiFi hotspots are configured to allow guests access to email and basic web surfing. Just one or two folks streaming videos over a shared connection can bring the network down for everyone in the RV park, and several folks online doing "normal" web surfing will not leave enough bandwidth available for anyone to stream video.

Many RV parks have gone to limiting how much bandwidth their guests can use daily so that everyone has a fair shot at using the resource, and others specifically do not allow streaming video.

Even if you find a park that doesn't specifically forbid streaming or limit internet usage, please be a good neighbor and don't hog all the bandwidth. If anything, limit nonessential high-bandwidth usage activities to off hours – such as overnight or while everyone is at work or out sightseeing.

159

Streaming Video – Netflix, YouTube, Hulu, Etc.

So often we hear from folks that their mobile internet needs won't be so demanding because they're not trying to work online or attend remote virtual classes, they *just* want to stream some movies and TV shows.

Unfortunately, streaming video is one of the most bandwidth-intensive things you can do on the internet.

An hour and half high-definition movie on Netflix can easily eat up **4.5GBs** of data! For perspective, if you're paying by the GB, such as overage data on an AT&T Mobile Share plan at $15/GB, that movie would cost you $67.50.

Suffice it to say, streaming lots of hi-res video content is just not going to be financially feasible for most – unless you happen to have a truly unlimited source of data.

If streaming is absolutely going to enhance your life, the expense and risk of buying a used grandfathered-in Verizon or AT&T iPad plan may be a worthwhile investment for you. Or you might be interested in an unlimited on-device data plan from a less-robust carrier, like T-Mobile or Sprint. Each offers unlimited data plans for on-phone usage (and a little bit usable for tethering) while on its owned network.

Here are some tips that might help a bit with incorporating some video streaming into your travels without breaking the bank.

Netflix Tips

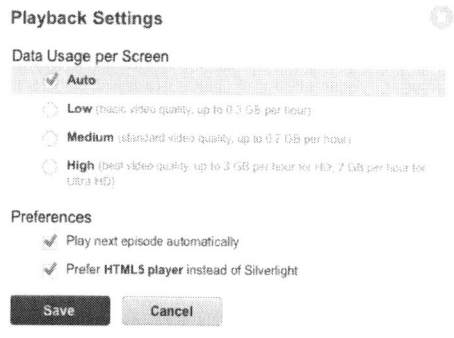

By default, Netflix will auto-adjust the quality of the video it is delivering to match your current connection speed. If you're on a solid 4G signal, this could mean very nice and smooth HD video, which will rapidly deplete data plan. You can force Netflix to show you content at lower quality levels so that you're using your data slower.

To select a setting that works best for your internet plan, navigate to the Your Account (movies.netflix.com/YourAccount) page and click Playback settings in the Your Profile section. Restricting data usage will affect video quality while watching Netflix – but on a smaller screen, it may not matter too much.

There are four data-usage settings to choose from; here's their estimated usage:

- Low (0.3GB per hour)

- Medium (SD: 0.7GB per hour)

- High (HD: 3GB per hour, Ultra HD 4K: 7GB per hour)

- Auto (adjusts automatically to deliver the highest possible quality, based on your current internet connection speed)

Switching to low-quality from high-definition video will allow you to watch 10 times the amount of video for the same bandwidth. But, of course, the video quality may be so low that it's not worth watching on larger screens or for big-budget movies with special effects. We find the Medium setting hits a sweet spot of data usage and quality for most our viewing.

YouTube Tips

YouTube will default to playing the best quality video for your connection speed but does provide a single setting to opt out of playing HD video.

Playback

Video playback quality

○ Always choose the best quality for my connection and player size
  Always play HD on fullscreen (when available)

◉ I have a slow connection. Never play higher-quality video

Go to your YouTube settings and select "Playback," and then select that you have a slow connection. This can help you automatically save some bandwidth, especially since YouTube auto-plays videos as soon as you hit a video's page.

For further control, you can install plug-ins into Firefox and Chrome browsers that force a specific definition of video for playback. For Firefox, check out "YouTube High Definition" and for Chrome check out "Auto HD." Both will allow you to downscale video playback as low as 240 pixels.

Roku Tips

Roku doesn't make it easy, but they have a hidden menu where you can adjust the quality of your video streaming.

For other video services that you might like to utilize, check their FAQs and user-settings pages to see if they offer a way to scale back the quality of the video they deliver to you.

Alternatives to Streaming Video

Instead of streaming content, here are some ideas for getting your media fix elsewhere.

- **Rent DVDs & Blu-rays** – Netflix isn't just for streaming! The discs-by-mail service actually works amazingly well for RVers who stop in places for at least a few days at a time, as you can update your shipping address frequently. If you're going to be staying for a few days in a location that has mail delivery, that's generally enough time to ship your current disc back and get a new one shipped to you. Netflix stocks a very wide selection of discs – including TV series and documentaries.

  RedBox kiosks are also very handy for renting new releases on disc for the evening, and they're located all over the place – Walmarts, 7-11s, drugstores, and more. You can rent a disk at one kiosk and return it to another down the road – all while only paying a low nightly fee, no membership required.

- **Rip content to hard drive** – While it's illegal to rip content to distribute, making a backup of it for your own personal use is in a legal gray area. We keep our legally obtained DVD collection in storage, but have ripped off-site backup copies of stuff we wanted to take with us to a hard drive. This gives us ample content to watch when we have no other way, without taking up valuable physical storage in our RV.

- **Download when you have WiFi** – When you have access to precious unlimited bandwidth, that is the time to stock up on content. We'll often buy a season of a favorite network TV show via iTunes and download it to have around. Always have extra hard-drive space handy for this!

- **Buy TV series on disc** – For series that we follow but don't care if we're watching the current season as it is aired, we'll buy the seasons on DVD/Blu-ray. Generally we buy used off of Amazon. When we're done, we'll sell them back online. This has become more difficult on Amazon.com, as they have put in additional restrictions for resellers – but we've found that selling a set of seasons on eBay can net decent results. You can also take them to pawn shops or used-media resellers, and turn them in for a little cash. Or exchange with fellow RVers for series they have. Note: You'll have to avoid spoilers for the current season from your family and friends on Facebook and Twitter.

- **EyeTV & DVR setup** – An EyeTV is a USB stick that acts as a TV tuner. Attach it to an external TV antenna and you can pick up local stations on your computer. It comes with software to turn your hard drive into a digital video recorder (DVR), so you can record content at specified times. This is an easy way to multi-task your computer setup, without investing in TVs, antennas, and separate DVR equipment.

- **Public TV Viewing** – Instead of trying to stream live events or major premieres of TV shows or sporting event, check around for local pubs that might be broadcasting. Sometimes it's fun to watch a major event with a group of fellow fans, and share a brew while you're at it!

Also keep in mind that the cost of paying by the GB to watch a movie might be less expensive and/or easier than going out to a movie or bringing in a rental disc (once you factor in time and fuel). Sometimes it's just worth using up spare bandwidth at the end of the month to treat yourself to some streamed content (ya know, use it or lose it!)

TV Antennas & Satellite TV

Many RVs come with a TV antenna built in that can pick up local stations wherever you roam. This is to referred to as OTA or Over the Air. You may find that this is good enough for keeping on top of local news, your favorite broadcast TV shows, and weather alerts. However, you will find a lot of variability in broadcast quality and variety, depending on how close you are to major towns and how strong of an antenna you have.

If televised content is important to you or you want more stations than is available OTA, you can subscribe to satellite dish services. Satellite has the advantage that if you can point your dish setup to the southern sky, you can watch television from wherever you are, you can get premium stations, and your channel numbering stays pretty consistent as you change locations.

Both Dish Network and DirectTV have options that you can take with you on the road – and which will work better for you depends upon your desired offerings for packages of channels, pricing, and contract terms.

Of particular note is how each handles local stations as you travel across the country. The major broadcast channels you receive will be dependent upon the service address on your account and if you're still within range of the beam serving that location. As you change locations, the providers will let you update your service address so you can get channels from your new location.

Dish also offers a DNS (Distant Network Service) that gives you major network programming that is not based in any one particular locale, if you prefer to not keep updating your address and perhaps depend on over-the-air TV antennas for local news.

You'll also have to consider what kind of hardware setup you want: portable or roof mounted.

- A portable system will allow you to park your RV in the shade under trees, and still get your satellite dish a clear view of the sky, or not worry as much if your campsite is perfectly aligned for satellite access. But this will require setup and takedown at each stop. If you don't plan to move around too often, this may be a great compromise.

- A roof-mounted system will require your RV be to be parked where your dish can see its satellite, which may not always be possible – but the setup significantly reduces your setup time, especially with an automated system that doesn't require manually retracting a dish. If you'll be moving around a lot, this has some definite advantages, as even overnighting will still give you satellite TV.

You'll also need to consider if you want a system that you have to manually aim each time you set up, which tends to be cheaper, or an automated system that can find the satellite on its own, which of course will be much more expensive.

There are lots of options for the equipment, and the prices will range from affordable to being quite an investment. Be sure to further research if you want to go this route to make sure you select the right setup for you.

Gaming on the Road

Online gaming demands a lot out of a network connection – particularly action games and first-person-shooters.

But to the surprise of many, online console and PC gaming over mobile data connections is actually very doable, if you are careful to manage your data consumption and overall expectations.

Online Gaming: Performance

Online games thrive on having a reliable low-latency connection, which is a measurement of the round-trip time between your computer and the gaming server. A home cable or DSL line usually has latencies well under 50ms, and this is what many games are designed for. A 4G/LTE connection can begin to get into this ballpark, with latencies between 50ms and 120ms. 3G connections are slower, but for some games the latency will still be plenty playable. Satellite, on the other hand, does not have game-friendly latencies. With lags of 1000ms or more, your character will be dead before you know what hit you.

A lot depends on the type of game. Massively multi-player online role-playing games (MMORPGs) and driving games tend to do really well with higher latency connections. Real-time strategy games are more challenging as latencies increase, as are first-person-shooters. The type of game that will really suffer even on an LTE connection are twitch fighting games that depend on precise timing for move combos. If you want to unleash that sort of pounding, you'll need to look for an alternative connection.

The other thing you will need to be prepared for are glitches and dropouts. A cellular connection is never going to be as reliable as a hard line, and inevitably some radio interference will someday happen that turns glorious victory into agonizing defeat.

Online Gaming: Configuration Issues

Some games will work without any special configurations – but just like with some home connections, some games may require special router configurations to be able to host multiplayer games. If you hit a dead-end, 3GStore.com actually offers a free gaming tip sheet and tech support to customers who have bought a mobile router from them, and they also sell this tip sheet directly to noncustomers.

Online Gaming: Data Usage

The biggest fear many gamers have is that online gaming will burn through their entire monthly data bucket in the blink of an eye. And this is a very legitimate fear!

But in general, most games are not data hogs. 3GStore.com tested a range of games, and found that most games burned between 50MB and 75MB an hour, even when using in-game voice chat features.

But you have to be very careful to keep on top of your data usage!

Games and gaming consoles are not designed to respect bandwidth caps, and it can be hard to keep usage under control. Many games will automatically install game updates and downloadable content, often without even giving any notice.

A downloadable content patch full of new levels may burn through gigabytes! One of our readers was able to use bandwidth-monitoring tools to catch his Xbox One in the act – waking itself up at 2 a.m. and downloading 4GB of data without asking for permission to do so, and with no notice given after the fact!

To be on the safe side, keep your gaming consoles disconnected (and unplugged!) when you are not actively using them, and keep a close eye on

165

your usage. And you may find that there are some games that are just too piggish to play on a capped mobile connection.

Online Gaming: Teams & Guilds

Being part of a raiding guild, clan, or team is going to be a challenge for mobile gamers. You do not want to be in the position of letting your friends down when you have such variable connectivity and always changing circumstances. As a mobile gamer, it is better to focus on games where you can play on your own time schedule, not tied to a team.

Online Gaming: Alternatives

Hardcore PC and console gamers often look down on mobile gamers – but there are actually some pretty amazing near-console-quality games out now for mobile devices. And the best part – unlike console games, most table games are optimized for the realities of mobile gaming. You may find that you can fully get your online gaming fix without ever firing up a console.

If you absolutely need your PC, PlayStation, or Xbox fix, seek out local gaming and/or comic book stores in your travels – many have gaming rooms with fast WiFi setup for gamers to come in and play.

This can be a great way to meet and play with local gamers and get some intense gaming in without needing to worry about latency or data caps.

Get Out!

You'll probably also encounter a good number of folks who tell you not to fret so much over your TV, movie, or gaming preferences.

After all, you're RVing and should just get outside. Why not explore and go on a hike?

If you're living on the road any substantial amount of time, it's about balance. If you enjoy a good movie or keeping up with your favorite shows or sports, there's no shame in that.

This is life on the road, not a vacation.

There will be bad weather days, or days you're not feeling well, or just days you're overwhelmed exploring another location. Unwinding after a day of work, exploring and socializing in front of the tube isn't a crime. Heck, we enjoy a well-put-together film – it's art!

But they do have a point – don't forget to get out there and explore too! If you're not able to get enough bandwidth today to stream your favorite show, maybe it is time to go out for a hike, read a book, or watch moss grow instead.

166

Beyond Data – Phones & Plans

A cell phone used to be for one thing – making calls.

And that was only if you were lucky enough to be in a coverage area.

Today, however, making calls is just another app on what has become an incredible pocket-sized technological Swiss Army knife.

For many people, a smartphone has replaced their camera, their GPS navigation system, their game console, their CD rack, and for some it has replaced their need for a separate computer entirely.

And, of course, for a lot of us, our phones have become our primary conduit to the internet – both directly and as a hotspot serving all our other devices.

With so much riding on one little device, it is little wonder that picking a phone is a big decision – and something that we get asked about often.

Choosing a smartphone is an intensely personal decision – and there is no universal right answer. Take some time to figure out what feels right – get some hands-on time with both the hardware and the operating system. Don't be swayed by splashy commercials, try not to focus on price, and absolutely do not lend any credence to the recommendations of store clerks.

You will need to make up your own mind, but here is some of our key advice to help steer you in the right direction.

Choosing a Smartphone: iOS, Android, or Other

Apple's iOS operating system powers iPhones and iPads. Google's Android powers most "smart" devices from other manufacturers – including

167

Samsung, LG, Motorola, HTC, and a whole slew of low-cost bottom dwellers.

Amazon is also using Android to power the Kindle Fire and Fire Phone, but Amazon has taken Android in such a divergent and incompatible direction from Google that it might as well be thought of as an entirely separate platform.

RIM (the maker of the old Blackberry) and Microsoft are still pushing smartphone platforms of their own, but they have had very little success winning over users and developers.

So at the moment it is a two-horse race, and for most people it comes down to choosing between iOS and Android.

Which to choose?

We get asked this question frequently – and for most users, our honest recommendation is simple: "If you have to ask and don't already have a preference, you are probably better off with an iPhone."

Android is a great platform for advanced users who like to play around with configuring and customizing the software on their phones. You can do some legitimately awesome things tweaking an Android device to fit you like a glove, things that are often impossible on an iPhone.

But if you don't want to make maintaining and tweaking your phone a hobby, and if you want something that works reliably and that will be supported for a long time – unless you already know you have a preference for the Android ways of doing things, most average users tend to be better off going with Apple's iOS.

But spend some time getting hands-on with both platforms and see what is most intuitive to you. They both have strengths and weaknesses.

Android has more options for customization and personalization. And with multiple manufacturers and devices, you have more hardware choices and cheaper price points.

Another big consideration is who you can go to for support on your device. If you have techno-savvy friends or family who are willing to help you out, sticking with their preferred platform ups the odds they'll be able to help out when you need some guidance. A huge benefit for the iPhone is Apple's Genius Bars in their stores across the country. You can attend classes and get free support in person for any Apple product you own.

There's no Android equivalent for device and OS support.

A note about subsidized devices: Particularly in the past, the standard way to buy a phone was to sign a two-year contract – and to then get that new phone for free, or $99, or $199. This is an incredibly enticing deal – particularly considering that the mobile carrier's true cost to buy you that phone might be $300–$500 more!

You get a huge discount up front, and the carrier makes back the cost (and more) over the course of your contract. To keep you from trying to duck out, hefty early termination fees apply.

This subsidized purchase price is all fine and dandy – if you take advantage of upgrading your phone as soon as your contract is up, or even hounding your carrier for early upgrades. (Store managers can often authorize this up to six months early!)

But if you still have a phone you bought on a two-year contract more than two years later, you are giving a huge gift every month to your mobile carrier. You are essentially continuing to make payments, even though your phone is already long paid off.

If your phone is over two years old and you are out of contract – STOP overpaying.

If you like your carrier, consider getting a new subsidized phone cheap and signing a new contract. You might very well be able to sell your old phone for more than this upgrade costs you.

Or keep your phone, and switch over to a cheaper plan with special rates for out-of-contract phones. Your carrier isn't going to do this for you automatically, but a phone call can save you substantially.

If you want a phone upgrade and do not want a contract, most carriers now offer attractive financing plans too.

But whatever you do, don't keep paying more each month for a phone you bought two or three or even four years ago!

169

Tips for Choosing an iPhone

Things are easy if you decide to go with iOS.

Because Apple's product line has been kept so simple, the choice is easy - buy whatever you can afford and what feels right in your hand.

Apple hardware tends to be really well built, and since the software is supported for so long, older phones tend to hold their value. If you aren't up for buying the latest and greatest, you can easily find a used phone that will still give years of good service.

Apple has a demonstrated track record of providing ongoing support for the hardware they sell – providing easy and free OS updates for years after launch, and free in-person tech support at Apple Stores too.

As we write this (February 2015), we'd consider the iPhone 4S the oldest iPhone worth buying, and the iPhone 5, 5C, and 5S, 6 and 6+ are all absolutely awesome options.

If you like to be on the cutting edge, you can buy a new iPhone every year and trade in or sell your old iPhone to keep the incremental costs under control.

Tips for Choosing an Android Phone

The Android operating system has begun to approach Apple's level of polish in the latest releases, but the quality of the hardware running Android is all over the map.

Some of the high-end Android flagships are every bit as solidly built as the iPhone – the HTC One M8 being a stellar example.

A lot of the lower-end Android devices, on the other hand, are so cheaply built that they are best avoided, no matter how low the selling price.

If you do go Android, please do not be tempted by these cheap generic phones or last year's now orphaned major-brand models. Invest in the latest Android flagship devices from Motorola, HTC, LG or Samsung and you will have the best possible experience going forward.

Make sure that whatever Android device you get ships with the current version of Google's Android operating system. Google uses candy-name code words in alphabetical order to indicate the version of Android – the recent ones have been named Ice Cream Sandwich, Jelly Bean, Kit Kat, and Lollipop.

Beware buying any Android phone that ships with Jelly Bean or earlier – that these phones have not been upgraded to Kit Kat by now is a near-certain sign that they never will be, and they have been orphaned by their manufacturer.

Android phone makers customize the operating system with their own unique look and feel, with bonus features and bundled apps of sometimes very dubious value. If you want to experience Android as Google intended, look for phones branded "Nexus" or "Google Play Edition." These devices ship with a stock build of the Android OS and are able to receive OS updates directly from Google.

If you don't have a device that is being directly supported by Google, you are at the mercy of your cellular carrier and device manufacturer for OS and security updates. And a lot of them have a very lousy track record of following through on updates – there are millions of Android phones out there with known major security holes that will never, ever receive fixes.

Only the major brand flagships tend to get much ongoing support, especially a year after initial launch.

Safe Shopping at Smartphone App Stores

Some of the best things about a smartphone, whether iOS or Android, are the amazingly powerful and creative third-party applications that you can install.

There are absolutely incredible applications available for both platforms. The tricky thing is finding the gems and avoiding the scams and malware.

Apple has a well-earned reputation for keeping malware out of the iOS App Store. The Google Play store, on the other hand, sometimes seems to be overwhelmed with scammers.

Be especially wary of "free" applications from brands you do not recognize that may be selling you out behind the scenes.

One recent example among thousands: The deceptively named "Brightest Flashlight" app (because no app can make the light any brighter than any other) that had been installed onto over 50 million Android phones was busted by the FTC for sneakily tracking and selling users' location details.

And if you think you can avoid the scams by running a security application – it was recently revealed that the former #1 paid app in the Google Play store, "Virus Shield," actually did absolutely nothing behind the scenes other than charge you $3.99 and provide a false sense of security!

When even the top-selling antivirus program is a scam, you know you've got a problem. Be careful out there!

171

Shopping for a Cellular Plan

Earlier in this book, we talked about cellular data plans – but here are some of the options to consider for adding voice service to the mix as well.

Each carrier has its own set of plans, ranging from prepaid plans that are ideal for short-term needs to long-term plans that may require a contract. Carriers are also constantly changing their offerings, as they evolve and compete with each other.

And then there are MVNOs (mobile virtual network operators), who have special arrangements with the major carriers to buy service wholesale and repackage it into deals worth exploring.

And regional carriers roam onto the larger nationwide carriers when traveling outside of your home network – meaning there are even more potential plans.

When comparing plans, here are some things to consider:

- Is there a contract required, and what is the penalty for breaking it or suspending it? You may find your travel plans change (such as taking an extended international excursion or a nomadic hiatus) or that new technology comes out that you want to switch to. Try to minimize the number of long-term contracts you get yourself tied to.

- Can you put your plan on "vacation" or suspend it? This comes in real handy if you find yourself staying in a spot for a while where you don't get enough signal to keep online, or where you might have access to other reliable options.

- Is the device you're buying subsidized by the carrier? Many devices are offered to consumers at steeply discounted prices – such as smartphones that may be offered for just $199 with a new contract, whereas the full price without contract is over $600.

- Does the data plan include hotspot/tethering, or is the data plan just for on-device usage? This incredibly convenient feature allows you to use your phone's data to connect your other devices.

- If you're shopping an MVNO reseller, which underlying carrier is the plan you're considering built on? This is the best way for you to be able to track down the real coverage map.

172

Overview of Current Plans Available

As these are things that change often, we're not going to focus on detailing all of the options here or the costs. But we'll highlight some of the most common cellular plans that we encounter fellow RVers utilizing in their travels.

Keep Current Alert:

We do track all major cellular data options in a price comparison guide for our Mobile Internet Aficionado premium members. If you opt to join, you'll get access to the constantly updated document to help make your shopping for a plan easier.

Mobile Share & Family Plans

Both AT&T and Verizon are now offering Mobile Share (AT&T) or the More Everything (Verizon) plans, which – if your family has multiple devices with the carrier – may make a lot of sense. With these plans, you get unlimited phone minutes and texting, and then a shared pile of data that is able to be used by all the devices on the plan, including for hotspotting.

And amazingly, the data packages you can purchase these days provide actual usable amount of data for folks on the road – up to 100GB per month.

You pay a fixed-per-device rate for each line of service you add to the plan, with discounts if you go contract free without subsidizing the cost of the device.

Other carriers offer family plans in which you can connect multiple phones together on the same account and get a discount; however, each phone has its own bucket of data – which may make better sense if you don't have people you trust to share and manage data with.

All of the carriers of course also offer individual plans. If you have trusted friends or family, it may still be worthwhile teaming up and getting a shared plan to reduce costs.

> **Hint:** Check back with your carrier periodically to see if they are offering new plans since you signed up. If there are newer and better deals available, you can oftentimes switch, but you have to ask. For instance, in October 2014 both AT&T and Verizon offered 'double data deals' for a limited time, but you had to request the new rates. Many informed consumers were able to snag 30-60GB plans for half price, and keep the pricing for as long as they like.
>
> (Keep tuned to the RV Mobile Internet News center at http://www.rvmobileinternet.com/blog - we report this stuff as we learn

173

about. You can subscribe to our RSS feed, or our free monthly newsletter too.)

Prepaid & Pay as You Go Plans

If you're just needing phone and data service sporadically, or on an as-needed basis, a prepaid plan may make the best sense for you. You can also set these plans up without contracts, activation fees, or credit checks. You just fill your account with credits or buy refill cards in major stores – and redeem the credit when you want to use your phone.

Rates vary from $35 to $60/month depending on the service you select and the network, and you can pick them up in stores across the country.

Verizon calls their service Prepaid, AT&T calls theirs GoPhone, and Sprint Prepaid.

The downside to prepaid is that you do have to remember to top up your account, and the available usage area may only be on the carrier's privately owned network, without access to roaming areas.

Some of the carriers also offer a Pay as You Go Plan, in which you just pay for the days you need to use your phone – as low as 99 cent to $1.99/day.

MVNOs

There are also resellers of the major networks that can offer comparable or better deals. Some of these companies offer plans on multiple carriers, and which service you end up with will specifically depend on the phone you choose. You'll need to do your homework when selecting the phone (or bringing your own) to know what carrier it was meant for.

Here's some of the popular MVNOs:

- **StraightTalk** (ww.straighttalk.com) – Offers a flat-rate plan of $45/month on either AT&T or Verizon. It's listed as unlimited talk, text, and data – however, data speeds get throttled after 3GB of usage a month, and the data is not tetherable.

- **Tracfone** (www.tracfone.com) – Offers flat-rate monthly phone plans on the major carriers, starting as low as $9.99, as well as pay-as-you-go services.

- **Consumer Cellular** (www.consumercellular.com) – Offering plans on AT&T, they have no-contract, build-your-own plans starting at just $10/month. Add on just the services you desire – from voice, text, and data.

- **Virgin Mobile** (www.virginmobileusa.com)- Offering plans on the Sprint network, you can get a plan with

174

voice, text, and unlimited data (limited to 2.5GB of high speed per month) for as little as $35/month. You can even turn on mobile hotspot for $5/day as needed.

Also available exclusively through Walmart, are the "Data Done Right" plans which include the mobile hotspot feature for free on multi-line plans, starting at $65/mo for two lines and 4GB of LTE data.

- **Net10Wireless** (www.net10wireless.com) – A division of Tracfone, they offer custom build-your-own plans on all four of the major carriers.

WiFi First Plans

A new type of MVNO has emerged that defaults to routing calls over WiFi; only if there is no WiFi available does it fall back to cellular service. This keeps costs way down – yet still gives you the ability to get online when out and about.

The pioneer in this model is Republic Wireless (www.republicwireless.com), which offers phone service starting at just $5/month for WiFi-only calling, $10/mo for unlimited talk and text on WiFi and cellular (Sprint's network), and $40/mo for unlimited talk, text, and 4G/LTE data (throttled after 5GB).

Republic also makes it easy to change plans frequently – as often as twice a month. If you are looking for affordable voice phone service, this can be a great option.

The biggest catch of Republic Wireless is that, since the phones need to be customized to default to routing voice calls over WiFi, only two phone models are currently supported.

Another similar WiFi First Sprint-based MVNO is Scratch Wireless (www.scratchwireless.com), which actually provides completely free monthly service for WiFi-only users, as well as free text messaging on cellular networks. If you do need voice or data when away from WiFi, you can buy daily ($1.99) and monthly ($14.99) passes when needed.

Carrier Redundancy

As with data, there is no perfect carrier. Although Verizon currently has the most coverage anywhere, other carriers have strengths in some markets – and it's always variable as to where you can actually get a signal.

Redundancy on multiple carriers can be a wise thing, especially for emergency use and if you have a lot of dependency on phone calls.

175

We discovered this one summer when we broke down in a rural area. The area we were in had absolutely no AT&T signal, which our voice capable devices were on. And the Verizon signal in the area was voice only – no data – so our Verizon Jetpack (data only device) could not get online.

But a kind stranger passing by happened to have a phone on Verizon that was able to get a voice signal, allowing us to get a call out to our emergency tow service for help.

We now travel with an emergency-only, Verizon-based, pay-as-you-go phone to increase our arsenal of connectivity. It is a small price to pay for piece of mind, particularly for traveling into remote rural areas where coverage for any given carrier can be very hit-and-miss.

Internet Calling Services

Not all voice calls have to be over cellular. Sometimes you may find that you have a good data connection by your MiFi device or a WiFi hotspot, but none of your phone devices have service.

There are several ways that you can make a voice call anyway.

- **Skype** (www.skype.com) – For free, you can place Skype-to-Skype voice and video calls, which require the other people you need to call to also have Skype. Skype is cross-platform and free to download to a computer or mobile device. It's pretty easy to use. And if you'd like to be able to call a phone number (mobile or landlines) from Skype, you can subscribe to Skype-Out service for as little as $2.99/month.

- **Google Voice** (voice.google.com) – For free, you can set up an account on Google Voice that gives you an online phone number that you can make and receive calls on. There are all sorts of configurations that allow incoming phone calls to be forwarded to specified cell phones and landlines, and you can even set quiet hours when calls go straight to voice mail. The voice mail system is pretty cool too – it translates voice to text, and will email and/or text the message to you. You can make calls via Google Chat from within Gmail.

- **FaceTime** – Only available between Apple devices (Macs and iOS), you can conduct video and voice sessions over a data connection.

There are also many other VOIP (voice over internet protocol) services out there, most of which are paid services and useful for business applications. Examples we have used include RingCentral and Line2.

Crossing International Borders

When we first hit the road, we thought for sure traveling through Canada and Mexico would be adventures we'd have in our first year.

Over eight years later, we still haven't crossed the borders in our RV.

And one of the main reasons has been the daunting task of doing the research to remain connected while we explored new lands. (The other big reason? We haven't run out of awesome yet in the USA!)

We've been putting the research off, and challenged ourselves a bit by proposing this new chapter during our crowdfunding campaign as one of our last stretch goals. Our supporters called us on it, so now we had to research it!

So as a big disclaimer on this chapter in particular – unlike other chapters – this is just research we've done. We've not actually put any of this to the test, although we have tried to confirm this information from multiple sources, including firsthand accounts.

General International Tips

The hurdle with international internet is not that other countries don't have plentiful options – it's getting connected to them as nonresidents who are just passing through the country on a short-term basis.

The tips offered in this section apply to both Canada and Mexico – as well as any other international travel you might embark on.

WiFi

Public WiFi is plentiful abroad, just like most other basic life necessities.

You should be able to connect in at campgrounds, coffee shops, cafes, libraries, hotels, airports, and more. When traveling outside the USA, it is likely going to be your cheapest and easiest solution, especially if you're only going to be in an area for a little bit of time when it may not be worthwhile tracking down other resources.

If you're planning to mix in international travels with your RVing lifestyle, assemble your technology arsenal to include options that can easily be portable and taken to hotspots. It's the world traveler's ticket to international travel.

As with all public WiFi hotspot solutions, the usual caveats apply. Expect intermittent speeds, being in crowded public spots, and keeping your own self secure. Using a virtual private network may be a smart idea for protecting yourself when regularly surfing from public portals.

Voice

There are several ways to keep your voice service working while traveling internationally. The simplest is to just activate international roaming with your home carrier. This way, your home number continues to work and will ring you wherever you go. You don't have to answer every call – but at least you'll know someone is trying to reach you, and you will have the option.

However, if you want to make or answer calls on a regular basis while you're out of the country, you'll want to avoid the default international rates your carrier is offering. Most carriers have special international packages that offer discounted calling rates and allow you to keep your number for as little as $15/month.

If keeping your domestic number isn't important, local SIM cards are abound – many with substantial discounts for calls to the USA.

And finally, you can rely on services like Google Voice, Google Hangouts and Skype to handle your calls.

Cellular Data

Our first tip is to turn OFF roaming on your devices when you are close to international borders – the roaming fees from your carrier can be astronomically high.

One of the quickest, albeit not cheapest, options is temporarily adding on an international roaming plan with your US-based carrier. This may be ideal if you're only planning a short trip, will primarily be relying on WiFi, or won't be needing much data to get by with.

Here are some details on the four major carriers:

178

Verizon (www.verizonwireless.com/wcms/global.html): For $4.99/month you can add international calling to your plan, which gives you a 20% discount off of per-minute rates, or in Canada/Mexico there is a $15/month option that gives you 1000 included minutes. And for data, if you turn on the feature, they'll charge you just $25 per 100MB of global data at 3G speeds (that's the equivalent of over $250/GB).

A special note for those protecting a grandfathered-in unlimited data plan – adding on the international plan requires switching plan types, which will discontinue your unlimited plan. And you cannot get it back after returning the States. Carefully consider this switch if your unlimited US data plan is important to you.

T-Mobile (www.t-mobile.com/optional-services/international.html) – Has a particularly attractive deal – some of their phone plans include free and unlimited international data in over 120 countries. And their voice plans offer a 20 cents per minute flat rate.

The caveat is, the data is going to be at EDGE speeds (128Kbps) – which isn't much better than dial-up. But it's still a pretty awesome and affordable deal for international data access.

If you need faster data, T-Mobile does offer packages starting at $15 for 100MB to $50 for 500MB to be used over a 2-week period.

AT&T (http://www.att.com/shop/wireless/international/roaming.html/att/global/affordable-world-packages): AT&T offers the Passport service, for as low as $30/month, which offers packages of discounted international voice calls, unlimited texting, WiFi hotspot usage and a little bit of high speed cellular data in over 150 countries starting at $30 for 120MB, or $120 for 800MB packages.

As of the beginning of 2015, AT&T is the only carrier which has international LTE roaming agreements, meaning that in many countries (including Canada) AT&T phones will connect at much faster speeds.

Sprint (www.sprint.com/travelingabroad): For $4.99 per month, Sprint will offer you discounted rates internationally, or for $2.99/month in Canada you can make calls within Canada and to the US, and receive them for only 20 cents a minute. For data, they offer international data for a multi-country trip starting at $40/month for 40MB and $30/month for 55MB in Canada alone.

If you're planning more extended time in a country, it may be worthwhile seeking out options with the local carriers to get a prepaid or no-contract cellular plan. Most of the rest of the world has standardized on GSM technology for their cellular networks, which is equivalent to what AT&T and T-Mobile use here in the States.

Having an unlocked global devices will enable you to pick up a SIM card in many countries you're traveling to, and by putting it in your device you can make calls and surf the internet from your own device at local rates.

Just make sure your device can use the frequency bands where you are traveling - many US devices are not compatible with the LTE bands used in much of the rest of the world and will at best give you 3G speeds!

If you don't want to do the research in each country you plan to visit, g3wireless.com sells a global SIM card that works worldwide. They have set rates for each country for talk time and data – however, it can become quite pricey. We checked prices in Canada and the data rate was 65 cents per megabyte. That equates to over $650 per GB. Not exactly an inexpensive way to go, but it does save a lot of hassle.

Another option is equipment rental company Telecom Square (http://mobilewifi.telecomsquare.us/plans.php) - for a flat daily rate starting at $12.95/day, you can get access to unlimited data in a variety of international locations. This is even available to Canadian citizens traveling into the US.

Canada

Canada is a very large country, with a lot of wild unpopulated areas. Internet just simply isn't as abundant as one might expect, and there are hurdles for US residents to get set up with their own independent source.

But with a little planning, you should be able to plan traveling in Canada with at least moderate internet usage. Anyone depending on lots of internet for their mobility may find it more difficult and expensive while traveling northward.

If you're working online from the road and wanting to traverse through Canada, you may want to carefully consider the costs and logistics for mobile internet (as well as temporary work visa vs. business visitor issues – but that's another topic). In our assessment, it might just be easier to visit Canada while in vacation mode.

WiFi

According to most RVers we've talked to who have ventured to Canada, WiFi seems to be present and usable in most campgrounds and RV parks – and many utilize them for their primary internet needs.

Aside from campgrounds, WiFi hotspots – like in most places across the world – are accessible from cafes, coffee shops, restaurants, libraries, breweries, and more.

Of course, public WiFi hotspots will vary from free to paid, and the quality of connection will vary highly. Having WiFi repeating gear on board can help improve the situation.

Cellular

Getting cellular service as a non-Canadian resident seems to be a bit tricky and expensive. To get a prepaid plan directly with one of the carriers, most require a Canadian credit card or banking account and address.

And once you have it, service outside of metropolitan areas can be spotty and unreliable. There's very little, if any, coverage in the wilds.

There are three major carriers in Canada – Telus (www.telus.com), Rogers (www.rogers.com), and Bell (www.bell.ca). Each has its strengths and weaknesses in different parts of Canada, so be sure to check their coverage maps for your travel plans before investing much energy into trying to get a plan set up with one.

> **For Telus –** you simply cannot get a plan as a US resident. However, if you have a friend who is a Canadian resident and a current Telus customer, the friend can add on a noncontract SIM card to their plan and you can reimburse them. We've read reports of others doing this with great success – the SIM cost around $12, and a 5GB plan cost just $40/month.

> **For Rogers –** US residents can purchase a prepaid plan through reseller Fido (www.fido.ca). There are resellers available in many malls and service centers. You'll need to purchase a device, which can cost up to $200 – and then a flexible monthly plan can be purchased that is billed as you need it (rates seem to be $10 for 150 MB, up to 5GB for $35, then $10 per additional GB).

> **For Bell –** US residents can purchase plans through reseller Virgin Mobile Canada (www.virginmobile.ca) by visiting kiosks in several chain stores like Walmart or The Source. As best we can tell from their website, data is only available as part of a cellular phone or tablet plan.

If you're sticking to metropolitan areas, you may be able to hook up some very affordable and fast prepaid mobile internet deals with www.Mobilicity.ca and www.Windmobile.ca – they offer service just in the five largest Canadian cities of Toronto, Montreal, Calgary, Edmonton, and Vancouver.

Another interesting option is Similicious (www.similicious.com/) – who takes the frustration out of searching for and activating a SIM as a non-

Canadian resident. They'll ship you a SIM card in advance or have one waiting at your first destination. They don't make promises, however, that the data on your plan will work as a hotspot, but they'll do their best to help you try. Rates for phone, texting, and data range from $45 to $85/month, and data only plans are $40 for 500MB to $70 for 2GB – so a bit pricey, but might be worth it to avoid needing to search for someone who will set you up with a SIM card.

The option of unlimited data through Telecom Square (http://mobilewifi.telecomsquare.us/plans.php) might also be an option worth looking into at daily rates of $12.95.

Canadian Bonus: For Canadians traveling in the US, here are some options worth considering:

US Prepaid Plans - Obtaining a pre-paid plan with the major US carrier might be feasible, but tricky. The carriers may require a US based credit card and mailing address. Between using a prepaid credit card, or picking up refill cards in stores - and using the mailing address of a US friend or relative - you might be able to navigate this.

KnowRoaming (http://www.knowroaming.com/)- Toronto-based SIM sticker maker is offering an unlimited data package for $7.99/day that covers customers in 55 countries they have roaming agreements with. You purchase a SIM sticker for GSM based phones and tablets for $29.99 that auto-detects when you leave your home country, and then authorize the charge while abroad. They can also supply a local phone number. We were not able to confirm if the service allows for personal hotspot or tethering of the data, or if it is on device only.

Roam Like Home (http://about.rogers.com/About/Media_Relations/News/14-11-07/ROAM_LIKE_HOME_in_the_U_S_for_5_per_day.aspx) - Effective November 2014, Canadian carrier Rogers launched a new service for their customers traveling to the US. For just $5/day (with a $50/month maximum charge), customers can access their plans in the US. So if a customer has an unlimited talk & text plan with a 6GB of shared monthly data - the entire plan can be used in the US too. Customers just need to text the word 'travel' to 222, to enroll in the plan and they are only billed for days they use the service.

Roam Mobility (http://www.roammobility.com/) - Utilizing T-Mobile's network in the US, Roam Mobility supplies a SIM card to Canadian customers for as little as $3.95 a day for 400MB of LTE data.

WIND Mobile (https://www.windmobile.ca/) - Newer Canadian carrier WIND Mobile began offering unlimited US roaming in 2014 as part of its regular plans, which may be an option for some.

Satellite

Traveling across borders is where satellite internet has a definite advantage. If you have a HughesNet satellite plan and the equipment on your RV, you can subscribe to a satellite that covers part of Canada and be connected anywhere you have access to the southern sky. There are no changes you need to make to your plan when you cross the border other than perhaps switching satellites, if needed. And the same rates and data limits you pay stateside apply while across the border. Satellites don't know about borders.

Mexico

Traveling to our southern neighbor seems to be a bit more accessible for bandwidth junkies, as long as they are willing to deal with slower speeds than they're used to in the States. We've heard of particular issues with obtaining enough fast internet for online working nomads in some of the more remote coastal areas like Baja.

WiFi

WiFi is available in some campgrounds, cafes, libraries, and public centers. Depending on the area, it may be slow and overloaded, particularly the farther from population centers you get.

Cellular

AT&T in early 2015 has made major strides in creating a North American cellular network by closing a deal to purchase Mexican cellular provider Iusucell. Once they complete the integration, AT&T's stated goal is to provide 'one network, one customer experience.' - which may make connectivity to and from Mexico a breeze.

But in the meantime, US citizens can easily pick up a Mexican SIM card from one of the two major carriers – Telcel and Movistar. If you're picking up a phone service in Mexico, it's very important to note that the country is divided into nine regions, and calling is considered long distance between them. So if you pick up a phone service in Region 2, your number will be assigned there. If you travel to another region, your number will now be roaming in the new region for local calls. If you'll be making in-country calls, it may be best to pick up SIMs for each region you plan to travel.

For data, however, there is no roaming when it comes to regions.

Telcel (www.telcel.mx) is the largest carrier in Mexico and has the most coverage, especially outside of major metro areas. Speeds are reported to be at 3G in most places.

You'll need an unlocked GSM device (UMTA/HSDPA 850/1900MHz), such as a device off of AT&T or T-Mobile, to utilize their service. You can preorder a SIM card while in the US, and then just pop it in your device when you cross the border – or you can stop in any Telcel office, which are widely available countrywide. According to other travelers, you may have difficulty setting up a new account at a kiosk station, unless you speak Spanish well or happen upon someone overly helpful. The offices tend to keep an English speaker on staff to assist visitors to the country. The SIM card (called "chips" in some places) will cost about 149 pesos.

Their Amigo service (www.telcel.com/portal/personas/amigo/detalles/ internet_amigo.html?mid=1107) is no contract and pay as you go.

You put money on account either online (use Google translate to change the screens from Spanish to English), or at any "recarga amigo" station (located in many grocery stores, convenience stores, or gas stations). And then you order your plan by sending SMS codes (the "Clave" column on the above linked table) to "5050."

You can activate plans that are good for 1 hour to 30 days, varying in usage limits from 10MB for 39 cents to 3GB for $33 (prices may fluctuate depending on exchange rates.). You can check your balance by texting *133#.

If you're going to be in mainly metropolitan areas in Mexico, then Movistar (www.movistar.mx) may be an option for you. They have less coverage throughout the country, but it is cheaper and faster. They have offices to set up your service in the markets they serve.

Satellite

Traveling across borders is where satellite internet has a definite advantage. If you have a HughesNet satellite plan, and the equipment on your RV, you can subscribe to a satellite that covers much of Mexico and be connected anywhere you have access to the southern sky. There are no changes you need to make to your plan when you cross the border other than perhaps switching satellites, if needed. And the same rates and data limits you pay stateside apply while across the border. Satellites don't care about borders.

General Recommendations

Everyone's situation is going to be unique. There's no easy way to make recommendations for what you should build into your arsenal without really assessing your needs.

But we can make general recommendations for some common scenarios.

The next pages will go over some sample needs for mobile internet that we've encountered amongst fellow RVers we've talked to – and the basic setup that we'd recommend to them.

Obviously, your mileage may vary and be influenced by factors unique to you, such as any contract you currently have in place or any specific usage needs.

None of these recommendation should be taken as a shopping list.

And of course – technology is changing all the time, and new options are constantly becoming available. So always do your homework first!

Mobile Internet Advisor

If you would like some personalized private advising, we do offer Mobile Internet Advisor as a service. We'll conduct a comprehensive interview with you about your specific needs, do a little research on your behalf, have a private phone or video chat session and then deliver a written report with our recommendations for you.

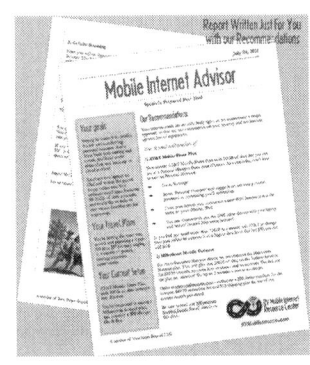

For more information, visit:
www.RVMobileInternet.com/advising

185

Lite Usage

Scenario: You just need to check email once or twice a day, do a little web browsing to research tomorrow's route and campground options, perhaps check into RVillage, and maybe an occasional video chat with loved ones back home. You're ok forgoing doing these things for a few days while you're off exploring a national park or out boondocking. Your internet needs are "nice to have" but not essential.

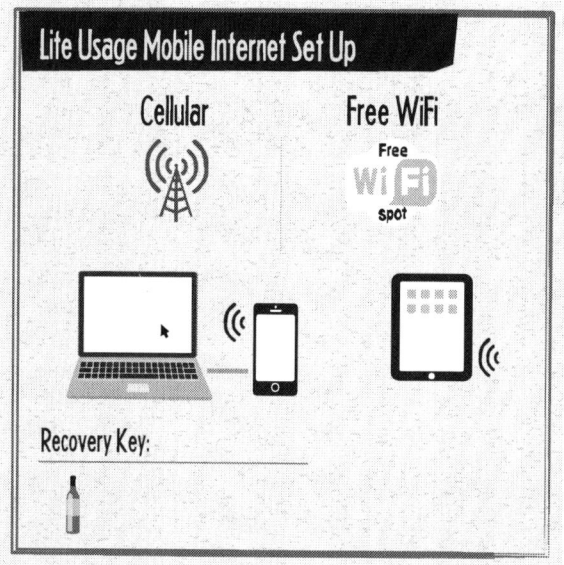

For this, having a tablet or smartphone device that can also hotspot to your primary laptop or computer may be sufficient. Get a plan with a smaller pot of data (maybe try 2–4 GB per month) to start, and scale as needed. You can supplement data from public WiFi spots when it's available.

If you have multiple folks in your RV-hold and each of them wants his or her own cellular device, then we'd recommend a shared plan, if available on your carrier of choice. Or you may find it worthwhile to have multiple carriers within your household – such as one member on Verizon and another on AT&T.

Or, additionally, you may find that getting a MiFi-type mobile hotspot device added to your plan for a dedicated internet source may be desirable. (You can typically add a hotspot to a shared plan for around $20/mo).

If you don't already have a preferred carrier in mind, we'd recommend Verizon as a starting place for most folks. Check their coverage maps, and see if they have service in the places you think you're most likely to go.

Moderate Usage

> **Scenario:** Keeping online is fairly important but not absolutely critical to your livelihood to be online all the time. But you'd really prefer to not go any considerable length of time without internet access.

You'll probably want an arsenal with at least two major connectivity sources, and those will depend on where you anticipate traveling, how much data you need, and your style of travel.

If you're planning to be mainly in populated areas with likely cellular signal and/or WiFi hotspots – two different cellular options on different carriers may make sense, with public WiFi being a backup and supplemental option.

If you don't already have contracts in place or preferred carriers and you'd like a decent bucket of data to use, we'd recommend getting a data plan on either Verizon or AT&T.

Moderate Usage Mobile Internet Set Up

Cellular

Free WiFi

Recovery Key:

You'll probably also benefit from a cradle-type cellular booster, like the weBoost Drive-4S – it's simple, affordable, and easy to use.

As always, WiFi hotspots can help fill in the gaps. If you find yourself in areas with public WiFi often enough, it may be worthwhile investing in a WiFi boosting system, such as the WiFiRanger Elite or Sky.

If you're planning lots of remote boondocking away from cellular towers, you may want to replace our recommendation for a high data cap cellular plan with satellite internet instead.

187

High and / or Consistent Usage

> **Scenario:** You need to be online fairly consistently, whether for work, school, or keeping in touch with family. You can probably survive being out of touch for a day or two.
>
> You'd like a setup that keeps you connected most of the time, and you realize that there will be compromises to make.

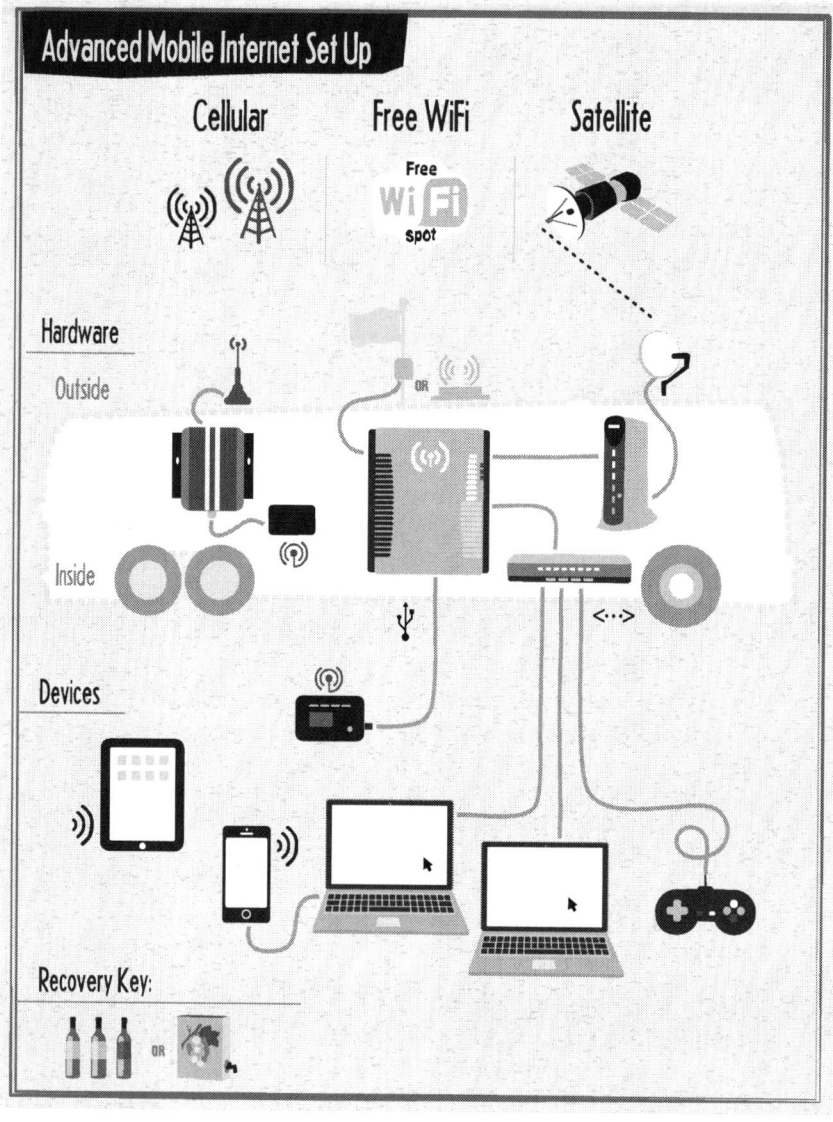

Advanced Mobile Internet Set Up

Cellular   Free WiFi   Satellite

Hardware
Outside
Inside

Devices

Recovery Key:

You'll want to design around redundancy and having multiple backup options. You'll also want high cap and/or unlimited data sources.

- We recommend Verizon as the primary cellular network you utilize. The Verizon coverage map is pretty extensive, and the LTE very fast.
  - If you already have an unlimited data plan with Verizon, keep it! Add on tethering/hotspot if you can. If you don't have one, perhaps consider 'renting' a plan off eBay if you're brave.
  - Pay Verizon directly on a Mobile Share plan for the data you want.

- We recommend a data plan on a second network to help round out your cellular footprint. In our experience, AT&T is the best complementary network to Verizon. Between the two, if there's cellular signal in the area – you're likely to get online. If you stream a lot of content on device, an unlimited plan on T-Mobile or Sprint might be very worthwhile too.

- To increase your ability to utilize cellular in more places, a robust cellular boosting setup is almost essential.

- A robust WiFi setup to bring in public hotspots can be a very worthwhile investment. We recommend a router setup (WiFiRanger or Pepwave SOHO) and WiFi antenna(s) on the roof of your RV. Remember, height is your friend – a mast of some sort with a directional antenna can vastly increase your range when stopped, and a more passive solution like the WiFiRanger Sky is great for shorter stops and closer-by hotspots.

- If you're planning travels outside of population centers where you may not have cellular signal or access to WiFi – a satellite may be worthwhile. Carefully consider if the costs and logistics are worth it to you, or if you can juggle your desires to be out in the boonies with your connectivity needs.

- For times you need intense data, specifically seek out RV parks with reliable WiFi networks and/or stay in monthly sites at parks where you can get cable or DSL installed to your site.

- Make space on board to carry lots of beer and wine to share with friends who have WiFi to share and space to park your RV.

  This is our category and you can see our personal current arsenal in the appendix of this book.

Absolutely Need to Be Online

Scenario: You absolutely, positively, must be online and connected no matter where you roam, with the fastest possible speeds. You need to handle huge file transfers on a consistent basis, and can't go without streaming movies and TV shows. 500-1000 GB of a data a month would be great.

And you want to go everywhere – urban camping to boondocking in the wilds far from civilization.

Constant and fast internet access is critical to your mobile livelihood and happiness in life.

You're in luck, there is an ideal technomadic vehicle for you: Air Force One.

If you are traveling on any less of a budget, you need to come to grips with making trade-offs between speed, cost, convenience, mobility and coverage.

Wrapping Up: Top 10 Tips

There is a LOT of information in this handbook. To help wrap things up, here are some of our best words of wisdom distilled down to their core.

Tip #1: Embrace multiple pipes!

The more possible on ramps to the internet at your disposal, the more likely you are to find one that works. Embracing a diversity of connection types and networks is the best possible way that you can maximize your chances of getting at least somewhat of a workable connection, particularly since WiFi alone is rarely going to be enough.

Tip #2: Soak up any WiFi you find!

Often the fastest, cheapest, and easiest way to get online is to use public WiFi networks, and in some parts of the country and world these are growing increasingly easy to find. Many libraries, coffee shops, RV parks, motels, and even fast food restaurants now offer free WiFi. Use it when you find it!!! But don't count on it - usably fast WiFi is often a rare treat.

Tip #3: Understand roaming & coverage issues!

One place where all the carriers are a bit deceptive is around roaming. Because they want their networks to seem as large as possible, they go out of their way to hide from you that you may be roaming and running into usage limits hidden deep in the fine print of their contracts. Stay on guard, especially if you are on Sprint or T-Mobile!

191

Tip #4: Be aware near borders!

Beware of international borders! All carriers charge an arm and a leg for international roaming (including onto the onboard cell networks offered on cruises now). If you are going to be anywhere close to an international border, make sure to turn off data roaming on all of your devices.

Tip #5: Know your caps!

Most fixed-location internet connections are unmetered, but mobile data is very commonly capped, and often comes with overage charges for excessive use. Save your big downloads for the days you have unlimited access.

Tip #6: Avoid contract lock-in!

Though you can often get your hardware cheap or free if you sign a two-year contract, these contracts come at a price of severely limiting your technomadic flexibility to change carriers or even countries on a whim.

It is often better to buy used equipment and avoid the lock-in if you can.

Tip #7: Learn parallelizing & batching!

Things like email and syncing RSS readers work wonderfully in the background on a slow connection. But web surfing can feel painfully slow if every new page takes minutes to render.

To deal with this, parallelize your browsing using multiple tabs. Whenever you see a link you want to follow, select "Open Link in New Tab" and make sure your browser is configured to load tabs in the background.

Tip #8: Boost what you have!

A signal booster and/or an extensible antenna mast can work wonders to help you get online from afar. These systems aren't magical, but on several occasions they have made the difference between having a barely detectable signal and a barely usable one.

Tip #9: Stay safe out there!

The internet is a scary place – and public networks can be especially so. To keep yourself safe, never ever use the same password in multiple locations. To take your security and privacy even further, subscribe to a VPN service.

Tip #10: Final Tip – We Repeat – Manage Your Expectations!

If you are planning in advance on having good net days and bad net days (and even no net days), you can better manage your own expectations around what you will be able to get done online, and when. Managing your expectations is perhaps the ultimate key to avoiding frustration!

Appendix A: Your Authors' Arsenals

Technomadia's Setup

As of the printing of this book (February 2015), this is the current setup that keeps us online most everywhere we go.

We're always tweaking this as new technology comes out, so be sure to check back with us here for any changes:

http://www.rvmobileinternet.com/resources/technomadias-mobile-internet-setup/

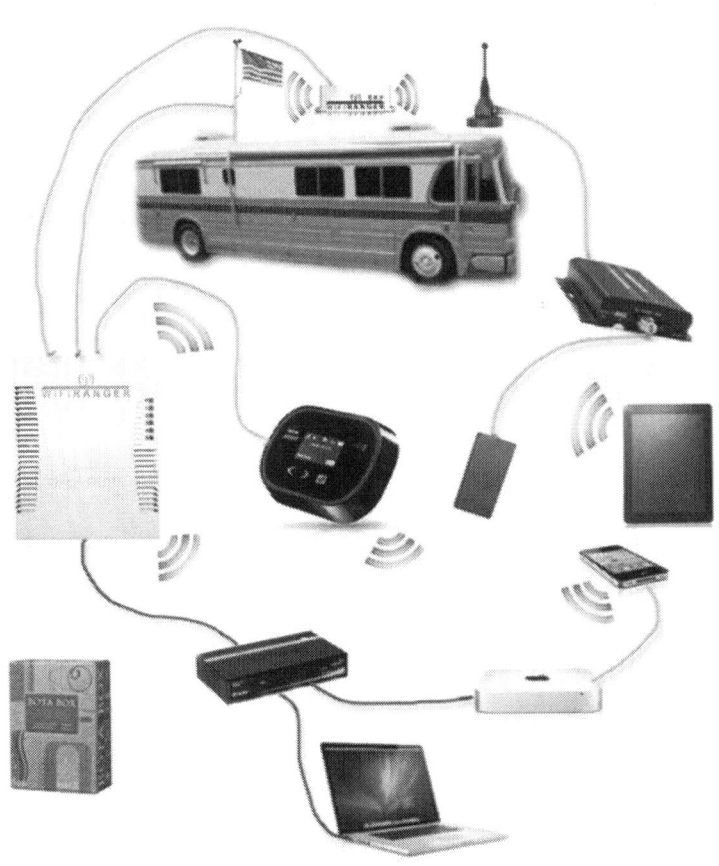

Cellular Options

- Novatel MiFi 6620L Mobile Hotspot with a grandfathered Verizon unlimited data plan ($69.99/month)

- 2 iPhones on an AT&T Mobile Share Value plan with 40GB/mo shareable (via Personal Hotspot) data ($180/month - obtained under a 'double data' promotion)

- iPad Mini on a grandfathered AT&T unlimited data plan (not shareable) ($29.99/month)

- iPad Mini with a T-Mobile "Free Data for Life" 200MB/mo plan (Free!)

At current time, none of our cellular options other than the T-Mobile free plan are available to new customers.

Antennas & Boosting

- **Mast:** When we're set up for a while somewhere, we put up our FlagPole Buddy 22' mast, which not only displays our colors but also is our mount for antennas we are testing.

- **WiFi:** We have a WiFiRanger Go2 router inside paired with a WiFiRanger MobileTi (no longer available) on the roof. This allows us to utilize WiFi hotspots from afar, and serves as the hub for our onboard network.

- **Extended Range WiFi:** We have a Ubiquiti NanoStation with a directional antenna we mount on the flagpole for further-away signals.

- **Cellular: Boosters:** Our current primary booster is the Wilson Mobile 4G (now called the weBoost Drive-4M), and we have been impressed with it. It is one of the first five-band, multi-device, FCC 2014 approved boosters to hit the market.

- If we need lots of bandwidth over a long period of time, we've been known to seek out a RV park were we can hook up our own cable or DSL internet.

Router

We use the WiFiRanger Go2 as our RV-wide router that brings everything together via both wired ethernet and private WiFi. We also use a generic gigabit ethernet switch to network our computers, NAS, and any other gear together.

Satellite

Up until November 2013, we traveled with a tripod satellite setup from HughesNet. However, between the prevalence of cellular and WiFi these days and no longer needing to be as on-call connected as we once were due to shifts in our client load, we ditched the setup after two years of nonuse.

Other

We keep a box of wine on hand for those times that nothing works. We've found that Bota Box RedVolution works best.

Jack Mayer's Setup

Be sure to check in with Jack at www.jackdanmayer.com/ communication.htm for any updates to his setup.

> **Router:** Our main router is the WiFiRanger Go2. This brings together cellular and WiFi, and enables me (Jack) to easily boost distant WiFi with a WiFiRanger Mobile on the roof. Other routers on hand that still are usable if required: Cradlepoint 95, 1000; Pepwave – all models of Pepwave back to five years ago…
>
> I bring all my communications gear together in one location and it is all wired for 12-volt use. My router runs all the time and we use it in motion. I also have a WiFiRanger X mounted in the truck in case we need it.
>
> **WiFi Capture:** A modified WiFiRanger Mobile is used most of the time, attached to the batwing TV antenna. This is modified to have an 8dBi omni antenna on it (like the WiFiRanger Marine or XT antenna upgrade). I also have a WiFiRanger Sky on the roof that I sometimes use.
>
> The batwing also mounts a Ubiquiti NanoStation M2 directional CPE. I have a dozen or more Ubiquiti Bullets I use for various purposes…and a couple of Ubiquiti PicoStations too. I've tested a Wave WiFi Rogue Wave WiFi capture device and occasionally use it - but my main capture is via the WiFiRanger.

Cellular: Samsung Galaxy S4 and Samsung Note 3, both on a Verizon shared-data plan and both hotspot capable. The Note 3 handles AWS spectrum – Verizon XLTE. We have our data "pool" set at 40 gigabytes. A Jetpack MHS291 mobile hotspot is used on Verizon, and also handles AWS XLTE spectrum. The Jetpack is attached to the WiFiRanger Go2 router wirelessly. I do not physically tether it.

Cellular Boosting: I have a bunch of legacy boosting gear for 3G that I no longer use. At the moment we use a Wilson Sleek 4G as our primary booster. Sometime in 2015 we will go to a new wireless boosting system that will better handle multiple devices.

Appendix B: Understanding 4G / LTE

Cellular data connectivity has been evolving rapidly over the past decade, with an entire alphabet soup of technical standards and protocols behind the scenes pushing more bits faster with each new generation.

It is more than most mortals should need to worry about – so the carriers adopted "G" for "generation" to give a simple marketing shorthand for comparing wireless speeds. When you see terms like 2G, 3G, and 4G, that's all it means – 2nd generation, 3rd generation, etc. In general, 3G is faster than 2G, 4G is faster than 3G, and…wait, what is this new LTE thing???

Here's a handy little infographic we created that will hopefully illustrate the evolution of cellular data technology a little better:

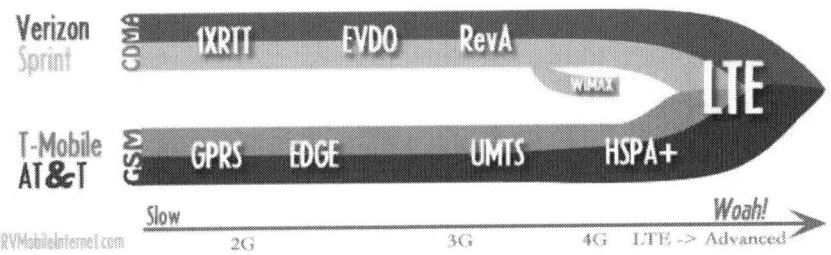

The Good Old Days

It used to be that there were two competing and fundamentally very different wireless technologies: In the USA, Sprint and Verizon used a technology called CDMA, and AT&T and T-Mobile and most of the rest of the world used an incompatible technology called GSM.

The third generation of CDMA technology was known as EVDO, and this was the first widely available wireless data standard that provided technomad-friendly speeds. When we hit the road in 2006, we relied on a Sprint 3G EVDO data card for years. It served us well, even though we were often actually happy to get 2G speeds in most places we roamed.

GSM networks were trailing at the time with 2G EDGE networks providing not much better than dial-up data speeds – fine for email, but painful for general surfing, and certainly not usable for video. But when the GSM networks evolved to 3G UMTS speeds, T-Mobile and AT&T leapfrogged ahead of CDMA's EVDO 3G speeds.

The future was bright for GSM networks AT&T and T-Mobile – with a clear technological evolutionary path mapped out from UMTS to HSPA+ (3G+) to LTE (4G), with the network growing ever faster and able to

197

handle increased user capacity. And as a global standard, the cost of LTE equipment would only get cheaper with time too.

CDMA networks, on the other hand, were at an evolutionary dead-end – there was no clear upgrade path for the carriers to anything beyond EVDO. Forging ahead would require a totally new investment in core cell-tower technology.

The 4G Revolution

Sprint bet big on a 4G technology called WiMAX and rushed to be the first to bring 4G service to market. Embracing WiMAX as a successor to EVDO seemed like a reasonable bet years ago, but now it has become a clear mistake and yet another dead-end.

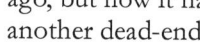

Verizon, on the other hand, predicted that the future was going to be LTE (aka Long-Term Evolution), and began aggressively building out the largest 4G LTE network in the United States.

Already being GSM based, AT&T should have had an easier time moving to LTE than Verizon, but AT&T let Verizon launch LTE before them and now AT&T has the second-largest domestic LTE network and is playing catch-up.

The following is a screen capture from our little app Coverage? that overlays all of the carrier's maps. We track these maps very closely on a regular basis. When comparing LTE to LTE, Verizon is clearly ahead:

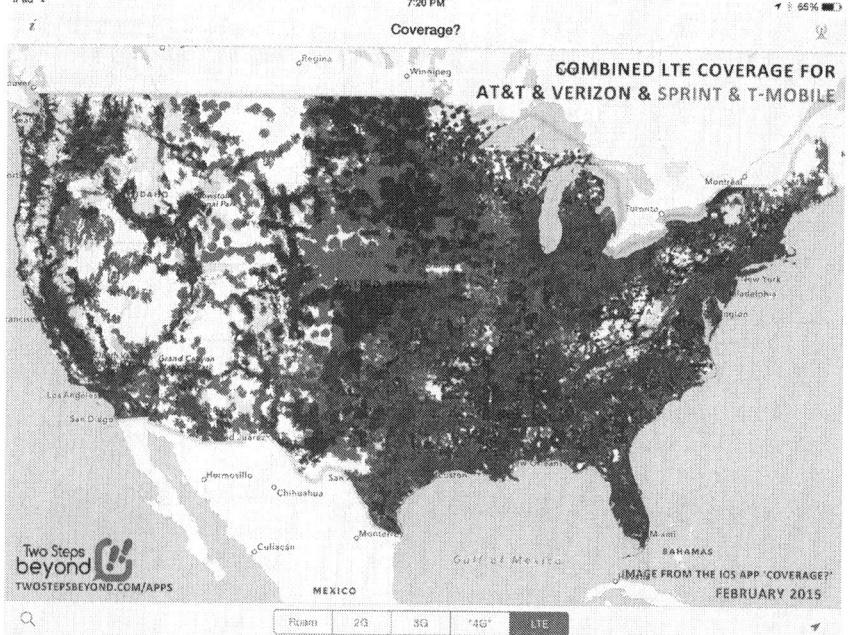

198

LTE Comparison: Verizon (red), AT&T (blue) and Sprint (yellow), a T-Mobile (purple). Verizon was clearly ahead in February 2015.

To combat Verizon's lead and 4G LTE marketing push, in 2012 both AT&T and T-Mobile decided to start marketing their HSPA 3G+ areas as "4G" – generating some criticism and lots of confusion.

But in practice, we've discovered that AT&T's 4G is always way faster than Verizon's 3G, and sometimes in practice actually rivals and occasionally even beats Verizon's LTE speeds.

And when you factor in this 4G coverage area, suddenly AT&T's network coverage jumps ahead of Verizon's, as shown below when we compare 4G to 4G:

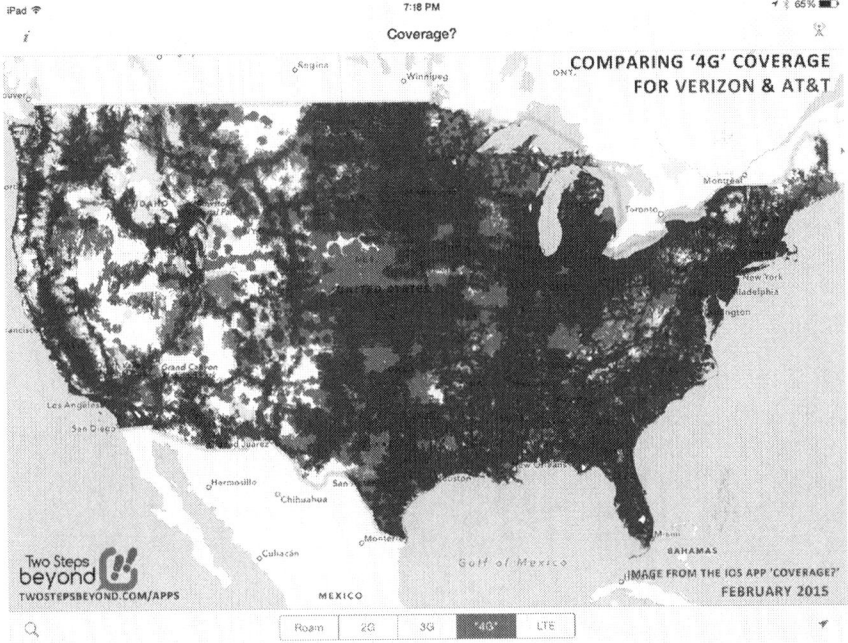

Seeing the LTE writing on the wall, Sprint has stopped expanding its 4G WiMAX network, and though they will keep it limping along until November 2015, buying a WiMAX device now is foolish.

Making up for lost time, Sprint has now started to roll out its own LTE network, but they are way behind both Verizon and AT&T.

T-Mobile is now also rapidly deploying LTE in markets across the country - at last bringing a common technological standard to all networks.

199

Are we on the verge of grand-unified LTE nirvana?

Not quite. The networking standards are converging to be at last the same, but the carriers own different and incompatible radio frequencies – and you can only fit so many different antennas into a smartphone. This is why there are often so many phone models – the Sprint/Verizon versions can speak to legacy CDMA networks and talk to the Sprint and Verizon LTE frequency bands, while the AT&T or T-Mobile versions may have radios designed to talk to incompatible LTE frequencies.

One other interesting twist: According to the International Telecommunications Union (ITU), HSPA+ and even LTE are not truly worthy of being called 4G technologies, just evolutions of third-generation technologies. According to the ITU, real 4G doesn't come until the rollout of LTE-Advanced over the next few years, when the networks will at last theoretically be able to deliver 100Mbps to mobile users, and gigabit (!!!) wireless speeds to fixed locations.

But regardless of whether or not the ITU considers it 4G, the marketeers have spoken – and both HSPA+ and LTE are being sold as 4G today.

What About Verizon's XLTE and Sprint's Spark Network?

These are both still LTE networks, though they are a further step towards LTE-Advanced, offering faster speeds and improved network capacity. AT&T and T-Mobile have also been evolving their networks as well – they just haven't come up with a clever marketing label for their evolutionary steps yet.

The Ecstasy & Agony of 4G

When you find yourself on an uncongested LTE network, it is an experience that can bring a bandwidth-hungry technomad tears of joy. Click, and pages open. Video plays without stuttering – even HD video. FaceTime video calls are almost like being there. Downloads are fast and painless.

At last, we have mobile data that is as fast as most people manage via their wired cable or DSL networks!

Even the upload speeds are fast!

But... LTE may be as fast as a home network connection, but only during ideal circumstances. And those circumstances can vary moment by moment, occasionally eliciting squeals of frustration.

We've seen LTE speeds all over the map in our speed tests, sometimes seeing vastly different results just moments apart. LTE speeds vary depending on signal strength, atmospheric conditions, how many other users are on the same tower you are connected to, and how much data is flowing upstream/downstream from that tower.

And it sometimes seems to even vary based upon the phases of the moon or the current hunger level of the cat – we just can't prove it yet.

Our experience with Verizon's LTE network has surprised us with just how inconsistent the performance can be – with speeds varying from a decent 5Mbs one moment to a blazing fast 20Mbps the next. We've even seen sub-3G 1Mbps speeds on occasion while showing five bars of strong LTE signal. It can be maddening at times, honestly.

One thing in particular that we noticed was how often LTE upload speeds actually clocked in faster than download speeds, probably because of the disproportionate load with so many more users surfing and streaming content.

In contrast, AT&T's 4G network has often surprised us with how consistent and reliable it is. While it may not offer as many peaks of blazing LTE speeds, it offers more coverage than Verizon's LTE and reasonable "good enough" speeds. We've had plenty of days where we actually end up preferring to switch our upstream connection to AT&T, even while in solid Verizon LTE territory. Surprisingly - we've actually run across places where AT&T's 4G performs better than AT&T's LTE!

We love having the ability to use whichever network works best at any given moment, and firmly stick with our recommendation that any technomad needing a near "always on" connection should rely on multiple networks for the greatest versatility.

Glossary of Terms

Before you get frustrated wondering why you might need a POE to power your CPE to get remote 802.11g when you'd really rather have more dB on your LTE – check through this glossary, and soon it will all make sense.

Even for the terms you know, or think you know, checking the definitions here might take your understanding to the next level.

2G/3G/4G/ "True 4G" The "G" suffix has become a handy way to label various generations of cellular networks – though the official meaning has grown murky as carriers have taken liberties calling faster versions of third generation networking technologies 4G.

Technically, none of today's LTE networks actually even meet the official definition of "4G" until LTE-Advanced is deployed. The International Telecommunications Union has dubbed LTE-Advanced as "True 4G" – and snobby network engineers call today's LTE networks 3.9G.

Here are the technical standards and the marketing labels in use by the four major carriers:

Verizon: 2G (1xRTT), 3G (EVDO), 4G/LTE (LTE, XLTE)

AT&T: 2G (GSM/GPRS/EDGE), 3G (UMTS), 4G (HSPA+), 4G LTE (LTE)

Sprint: 2G (1xRTT), 3G (EVDO), 4G (LTE, WiMAX, Sprint Spark)

T-Mobile: 2G (GSM/GPRS/EDGE), 3G (UMTS), 4G (HSPA+), 4G LTE (LTE, Wideband LTE)

802.11 The 802.11 standard is the formal name for the technology commonly referred to as WiFi. The 802.11 standard comes in

a/b/g/n/ac several flavors now:

- 802.11b – The first widespread standard. 2.4GHz, 11Mbps max.

- 802.11g – Faster. 2.4GHz, 54Mbps max.

- 802.11a – Operates on 5GHz, 54Mbps max.

802.11 **a/b/g/n/ac** **(continued)**	• 802.11n – Operates on 2.4GHz and optionally on 5GHz. By combining multiple MIMO antennas, speeds up to 600Mbps are possible. • 802.11ac – The newest WiFi standard – operates on 5GHz and delivers blazing fast speeds up to 1.3Gbps over short distances, such as around a house.
AirCard	The brand name for cellular modems from Sierra Wireless, but has become a generic term for modems that allows a computer to connect to a cellular network.
Amplifier	Pumps up the volume of a wireless signal. See also: "Booster" and "Repeater"
AMPS	Stands for Advanced Mobile Phone System, the original 1G analog cellular network widely deployed in the 1980s and 1990s. The AMPS network was phased out in the USA in 2008.
Analog	Analog radio signals are made up of waves that have not been digitized – meaning that anyone with a compatible tuner can listen in. Analog radio is easy to eavesdrop upon, is a very inefficient use of spectrum, and it suffers from static and other interference – especially as distance increases. The analog AMPS cellular network was at last fully shut down in 2008, but other wireless analog communication technologies remain – such as CB and ham radio.
Android	Google's smartphone operating system.
Antenna	An antenna takes electrical input and broadcasts it out as radio waves, or receives radio waves and provides an electrical signal out. An antenna needs to be designed and carefully tuned for the frequencies that it needs to support. An antenna optimized for traditional cellular networks may be useless receiving LTE signals. See also: "Frequency"
Attenuation	Attenuation is the measurement of signal loss over distance in a wire or through the air, usually reported as dB/cm or dB/ft. This measurement is important for antenna cables – "low-loss" cables have much less attenuation over a given distance. See also: "Gain"

Band (Radio)   The radio spectrum is divided up into bands based upon the signal frequency – ranging from the very low frequency bands used to communicate with submerged submarines to the extremely high frequency and beyond bands used for radio astronomy and exotic sensors and weapons.

Cellular networks all operate in the UHF (ultra-high frequency) band – which ranges from 300MHz to 3000MHz. Digital TV also operates in the UHF band.

Satellite communications and 5GHz WiFi are in the SHF (super-high frequency) band, which is defined as 3GHz–30GHz.

Standardized batches of LTE frequencies are also known as LTE bands.

See also: "Spectrum" and "Frequency"

Bandwidth   All of us mobile internet junkies know that we crave bandwidth, but how many of us actually know what the technical definition of "bandwidth" is?

Any wireless broadcast actually extends across a range of frequencies – say, for example, the old analog TV channel 2, which spanned 54–60MHz.

The bandwidth is the size of that channel – in the case of analog TV in the USA, it was always 6MHz per channel. FM radio, in contrast, is 200kHz and AM radio 10kHz.

The larger the bandwidth allotted to a channel, the more information that the channel can carry. But with larger bandwidth channels set aside, there can be fewer of them without overlapping.

LTE supports a range of channel sizes – with bandwidths ranging from 1.4MHz to 20MHz wide.

In general, an LTE tower can support 200 active full-speed users per 5MHz of spectrum allocated. Once those 200 slots are filled, network speeds start to fall for everyone. This is why unlimited data is so hard to offer on cellular networks.

(continued…)

205

Bandwidth

(continued)

LTE-Advanced will support bandwidths up to 100MHz, accomplished by combing multiple separate channels together. This will allow for many more simultaneous users and much faster peak speeds.

Prime ranges of frequencies suitable for cellular data use are in very limited supply, and the major carriers have paid billions to buy up as much bandwidth as they can. Very few carriers own enough spectrum to offer 20MHz LTE channels in many places, but this is changing. 5MHz or less is actually much more common.

What Verizon calls XLTE is an example of a 20MHz LTE network.

This variation in available bandwidth is one of the reasons why not all LTE networks are created equal.

See also: "Frequency," "Refarming," and "XLTE"

Bandwidth Cap

Also know as a data cap, this is an imposed limit on the amount of data that can be used over a certain period of time. Internet service providers sometimes refer to their implementation of bandwidth caps as a "Fair Access Policy."

Bars

On mobile devices, bars are a visual representation of RSSI, the Received Signal Strength Indication – an arbitrary mapping of signal power received (in dBm) to a number.

This internally calculated number is usually then mapped to a simplified visual display – bars or dots.

Apple now has dots, most cellular devices have bars, and WiFi devices come up with their own shapes all the time.

Also, bars are a great place to go to drown your sorrows when you can't get online any other way. As an alternative, consider keeping a box of wine handy.

Bird-on-a-Wire (BOW)

Satellite internet dishes need to be precisely aimed at the satellite providing data service. Rather than require a second dish for television reception, a BOW mounts additional receivers precisely offset on the same dish being used for data to also allow that dish to pick up satellite TV service too.

Bluetooth   Bluetooth is a slow, short range wireless communication technology most commonly used for wireless headsets and speakers. Though Bluetooth uses the same 2.4GHz frequencies as WiFi, it is designed to avoid interference - and usually succeeds.

Booster   Cellular signal boosters are amplifiers used to improve signal reception. Typically they have connections for an inside and outside antenna, amplifying the signal from the external antenna and rebroadcasting it indoors.

bps (Speed Measurement)   Stands for bits per second. The smallest unit of computing is a bit – 0 or 1. Network speeds are measured in how many of these bits are capable of being sent per second. It takes 8 bits to make a "byte," and one byte usually represents a single character of text.

Broadband   In radio communications, broadband refers to a higher bandwidth transmission capable of transporting multiple signals simultaneously. This term became popular in the 1990s for describing any internet access technology faster than 56Kbps, which was the speed of a dial-up modem.

Broadband is relative term – and over time what it takes to be considered "broadband" has been gradually increasing.

As of early 2015 the FCC has redefined its official definition to require speeds of 25Mbos down and 3Mbps up for an internet provider to be able to claim that they are offering "broadband" service.

Cable Internet/ Modem   Internet service provided by a cable TV company, over the same physical network of wires that provides TV service. Cable internet can provide extremely fast service.

Carrier   A cellular company is often referred to as a carrier because they carry transmissions on behalf of others. The legal status of being a "common carrier" means that cellular companies are not legally responsible for monitoring the content being passed over their networks. The liability for anything illegal falls upon the person doing the transmitting.

See also: "Operator"

207

Carrier Aggregation

One of the biggest challenges facing mobile network operators is that they have lots of 5MHz and 10MHz bandwidth chunks of spectrum, but to offer the fastest possible speeds LTE requires 20MHz, and LTE-Advanced can take advantage of up to 100MHz.

Carrier aggregation solves the problem by combining discontiguous channels from different chunks of spectrum to make a higher bandwidth virtual channel that can support faster speeds.

This is one of the core features of LTE-Advanced, but some carriers and devices began to support this on current LTE networks in late 2014.

See also: "LTE Advanced," "Bandwidth"

CDMA

Code Division Multiple Access (CDMA) is the common name for the IS-95 and CDMA2000 standards for 2G and 3G networks embraced by Verizon and Sprint in the USA.

Unlike GSM, CDMA is not widely deployed around the world and is thus not well suited for international roaming. Most CDMA World Phones actually switch into a GSM mode while roaming.

Going forward, legacy CDMA networks are being replaced by LTE.

See also: "GSM"

Cell/Cellular

A cellular network is based on dividing the coverage area into numerous slightly overlapping cells. A base station (aka cell tower) within each cell provides coverage to all the devices inside the cell and hands off responsibility to a neighboring cell when the user moves.

Cells can be as small as a room, or as large as 22+ miles (35+ km) in diameter. In congested or urban areas, smaller cells are used to allow for more people to use the network simultaneously.

Channel (Radio)

A channel is the combination of the frequency a signal is being broadcast upon and the bandwidth that it occupies.

In other words, the frequency is the address of the channel, and the bandwidth is the size of the house.

The more bandwidth a given channel takes up, the fewer of them you can have in a given slice of spectrum.

See also: "Bandwidth," "Frequency," and "Spectrum"

CPE

CPE stands for "customer premises equipment" and is the term used for commercial-grade WiFi access points used by wireless service providers. Very often, the equipment actually providing WiFi in a campground is a Ubiquiti or EnGenius brand CPE.

Advanced users can also use a CPE to maximize their own WiFi range.

The WiFiRanger roof mounted Elite, Sky, and MobileTi, The Wirie AP+, and the Wave WiFi Rogue Wave products are all actually commercial-grade CPE's under the hood with user friendly simplified interfaces tacked on top.

Decibel (dB)

The logarithmic decibel scale is used to indicate the amount of gain an amplifier or antenna provides, or the loss introduced over wires. Positive numbers represent amplification, and negative numbers represent loss.

On the decibel scale, every 3dB represents a 2x increase, and every −3dB decrease represents 1/2. 10dB represents a 10x increase, 20dB represents 100x, 30dB represents 1000x, and so on.

The FCC-limited maximum gain for a mobile cellular boost is 50dB – this represents an increase in signal of 100,000 times.

Decibel-Milliwatts (dBm)

It is very common to see signal strength expressed as a negative number with the units "dBm," which stands for decibel-milliwatts. But what do these numbers mean, and how should you compare them? And why are they negative?

On the dBm scale, a value of zero indicates 1 milliwatt of power. The decibel is a logarithmic scale, so every 3dBm increase represents a doubling of power, and every 10dBm represents a 10x increase. So 3dBm represents 2mW (milliwatts), and 10dBm represents 10mW, and and 20dBm represents 100mW, and so on.

The pattern works for negative numbers too. -3dBm is half a milliwatt, and -10dBm is 1/10th of a milliwatt.

By the time a radio signal reaches a receiver, the amount of power left is incredibly small – and thus is measured in negative dBm. A great signal for a wireless device would be a -50dBm signal strength. A barely useable signal is -100dBm, which is 100,000 times weaker. And even the most sensitive receivers will have a hard time picking up a -110dBm cellular signal, which is a million times weaker than a -50dBm signal.

It is amazing that such sensitive technology exists and can be packed into your pocket. But GPS radios are even more amazing – the typical signal received from a GPS satellite is around -127dBm!

DHCP

DHCP (Dynamic Host Configuration Protocol) is the magic that happens behind the scenes that allows your computer and other devices to automatically configure to connect to the upstream router. The router uses DHCP to assign each device an IP address and tells it what DNS service to use. If your computer and the DHCP server get out of sync, learning how to tell your computer to "Renew DHCP Lease" might be what it takes to fix things. (Rebooting accomplishes the same thing too.)

Dial-Up

Back in the dark ages, people used a telephone modem to dial a phone number of an ISP to connect to the internet.

Digital

Digital signals are made up of zeros and ones. All cellular networks today are digital networks. A digital network suffers drop-outs with distance, until eventually the signal is lost.

See also: "Analog"

Dish

The saucer-shaped antenna that focuses signals from a satellite onto the low-noise block (LNB).

210

DNS

DNS stands for "Domain Name Service" – This is the service that translates a name like "google.com" into an IP address like "206.181.8.251." Think of it like the white pages for the internet.

DSL

Stands for "Digital Subscriber Line" – home internet service provided via telephone wires, usually by a local phone company.

Ethernet

Wired networks are commonly called ethernet networks. Ethernet wires look like phone wires, though the jacks at the end are wider. The ethernet wire is sometimes called Cat-5.

FaceTime

Apple's proprietary ultra-simple video phone technology. With FaceTime, you can make a video call to other iPhone and iPad users, as well as to Mac laptops and desktops. But PCs and Android devices cannot make or receive FaceTime calls.

Frequency

Frequency is defined as the number of times a repeating event happens per second, and is measured in Hertz (or Hz).

A typical cellular radio might operated at 700MHz – which means that the peak of the radio wave passes 700 million times per second.

When it comes to radio waves, lower frequencies travel farther and can more easily pass through walls and obstructions. Higher frequencies are more easily blocked and are better suited for shorter range use. The advantage of higher frequencies is that there is much more bandwidth available – making it easier to increase network capacity and speeds.

To keep radios from broadcasting on top of each other, most of the radio spectrum is managed by the government and only licensed broadcasters can transmit on a ranges of frequencies that they "own."

The primary cellular frequency bands in use in the USA:
- 700MHz – AT&T and Verizon LTE.
- 800MHz – Cellular
- 1700/2100MHz - AWS
- 1900MHz – PCS

The primary WiFi frequencies in use:
- 2.4GHz – 802.11b, 802.11g, 802.11n
- 5GHz – 802.11n, 802.11a, 802.11ac

See also: "Bandwidth," "Spectrum"

Gain

The gain is the increase in signal provided by an amplifier or an antenna. Gain is logarithmic and reported in dB – a 3dB gain is a doubling in signal power. A 10dB gain is a 10x increase in signal power.

See also: "Attenuation"

Gbps (Speed Measurement)

Gbps stands for gigabit per second and is also often written as Gb/s. A gigagbit is made up of 1,000 megabits.

Computers often have "Gigabit Ethernet" wired networking ports, but gigabit ethernet routers remain somewhat rare.

The new 802.11ac WiFi standard paves the way for gigabit WiFi over short distances, and the evolution of LTE into LTE-Advanced paves the way for gigabit cellular connections!

See also: "bps" and "Mbps"

Ground Plane

Many antennas are designed to need a ground plane to function properly, especially the common magnetic-mount antennas many cellular boosters come with.

A ground plane is created by providing a metal surface underneath the antenna. Steel works great (like a car roof), and the magnetic antennas stick beautifully to roofs like this. But aluminum also works as a ground plane – you just need to find a different way to hold the antenna secure to the nonmagnetic surface.

A rubber or fiberglass roof is useless as a ground plane, but you can instead use a small metal disk attached with adhesive to create one. A 3.5" diameter disk would be minimum, and 8" or more is better. Some people actually use a cookie sheet!

GSM

GSM stands for "Global System for Mobile Communications" and is an international standard for 2G cellular networks. In the USA, AT&T and T-Mobile built their networks on the GSM standard.

Central to the GSM standard is the SIM card, making it relatively easy to move service from one device to another. GSM phones are often able to roam onto other GSM networks around the world.

Going forward, legacy GSM networks are being replaced by LTE.

212

Hangout   Google's group video chat technology.

See also: "Skype" and "FaceTime"

Hotspot   A hotspot is the common name for a WiFi network. A public hotspot allows those nearby to connect without needing a password. A "personal hotspot" is the term often used to describe a private hotspot made by a smartphone or MiFi to allow other authorized devices to share a cellular connection and get online.

iOS   The operating system inside Apple's iPhone and iPad.

IP Address   Every site on the internet has an IP address. Behind the scenes, this numerical address is used to communicate instead of friendly names like technomadia.com.

See also: "DNS"

IPv6   The internet is running out of IP addresses to assign to newly connected devices. IPv6 is the next generation of networking protocols that pave the way for trillions of connected devices and beyond.

LTE is designed from the ground up based around IPv6.

ISP (aka Internet Service Provider)   Who is providing your internet service? That company is your ISP.

Jetpack   Verizon refers to mobile hotspot devices as Jetpacks, and the term is often used interchangeably with MiFi.

Kbps (Speed Measurement)   Stands for kilobit per second and is also often written as kb/s. A kilobit is made up of 1000 bits, or 125 bytes (text characters). Old telephone modems were capable of receiving 56Kbps speeds – painfully slow by today's standards.

2G wireless networks are typically measured in Kbps.

See also: "bps" and "Mbps"

LAA-LTE   Currently an experimental technology, License Assisted Access (LAA) will allow carriers to deploy super-fast LTE access points that supplement their regular licensed LTE frequencies with unlicensed 5GHz spectrum, co-existing with 5GHz WiFi.

T-Mobile has indicated they may deploy LAA by the end of 2015, and Verizon is experimenting too.

LAN

LAN stands for Local Area Network, as opposed to WAN, which stands for Wide Area Network. Whether wired or wireless, the network inside your RV is considered a LAN.

Latency

When it comes to networks, latency refers to how long it takes for a remote server to respond to a request. The latency is often referred to as "ping time" by speed-testing apps and is reported in milliseconds.

Think of it as follows: When you click a link, how long before you see a new page start to load? That time delay is latency.

The average latency for wired home internet services in around 30ms.

On a good LTE connection, ping times of 75ms to 100ms are common, and this feels plenty fast in use. HSPA+ 4G and 3G networks are slower – averaging around 120ms to 170ms.

Satellite networks, on the other hand, unavoidably have massive latency resulting from the signal's speed-of-light round-trip to geosynchronous orbit and back. The best satellite systems achieve a latency of 600ms, and over 1000ms (one full second!) is not unusual.

A high-latency connection may be capable of fast raw transfer speeds, but the connection will feel slow for interactive use because of all the delays. Latency is particularly painful for audio and video chat applications (where the half-second delay leads to talking on top of each other), and for typing into remote terminal sessions.

And online action games are a really bad idea without a low-latency connection – otherwise, you will very literally be dead before you know it!

LNB

On a satellite TV or internet system, the LNB is the extremely sensitive reception antenna mounted on a boom at the focal point of the dish.

LNB stands for low-noise block.

LTE

The dominant 4G technology is known as LTE, which stands for Long-Term Evolution. The LTE standard is gradually supplanting the previous cellular technology standards that came before it.

LTE-A (aka LTE-Advanced)

The LTE platform was designed to evolve to support faster and more advanced networks. The next major step in the evolution of LTE is being called LTE-A or LTE-Advanced.

One of the primary advances coming into LTE-A is support for channel bandwidths up to 100MHz, 5x the current 20MHz maximum. To achieve this, LTE-A will support combining multiple separate physical channels into a single larger virtual channel.

The design goal for LTE-A is to enable cellular networks to support speeds up to 100Mbps for mobile users, and up to 1Gbps for stationary users.

Just a few years ago, these speeds would have been considered fantastical. Soon they will be available in pockets everywhere.

See also: "Bandwidth"

LTE Broadcast (aka LTE Multicast)

LTE Broadcast is a technology that allows for a video stream to be broadcast over LTE to multiple compatible simultaneous receivers - useful for live TV service and at sporting events.

Verizon has announced intentions to deploy LTE Broadcast service in 2015.

MB/s (Speed Measurement)

MB/s stands for mega*bytes* per second – be very careful not to confuse this with Mbps (mega*bits* per second)!

One MB/s is 8Mbps – and even computer savvy people often screw up and use the wrong abbreviation.

Networks are always measured in bits per second – Kbps, Mbps, and Gbps. But computer interfaces like USB ports and peripherals like hard drives often have their speeds described in megabytes per second – MB/s.

A common USB 2.0 port maxes out at 60MB/s, USB 3.0 hits 625MB/s, and Apple's Thunderbolt port can transfer 1250MB/s.

Mbps (Speed Measurement) Mbps stands for megabit per second and is also often written as Mb/s (not to be confused with MB/s). A megabit is made up of 1,000,000 bits, or 125,000 bytes (text characters).

Ethernet-wired networks used to operate at 10Mbps, and now 100Mbps "fast ethernet" is common.

3G and 4G wireless networks are typically measured in Mbps, as are WiFi network speeds.

See also: "bps" and "Kbps"

MiFi "MiFi" is Novatel's trademarked brand name for its line of mobile hotspots, but the term is often used generically to refer to any similar device.

MIMO MIMO stands for multiple-input, multiple-output – or, in other words, using multiple antennas working together to increase data speeds. MIMO technology is central to both 802.11n WiFi and LTE cellular.

Mobile Hotspot A mobile hotspot is a small device, usually battery powered, which combines a router with a cellular modem - allowing the user to share a cellular connection with other nearby WiFi devices.

Modem Modem is short for "modulate/demodulate" – the technical terms for transforming digital data to be transferred over a wire or a radio. The term "modem" was popular in the old dial-up days and is also commonly used to reference a cable modem or DSL modem.

MVNO Mobile Virtual Network Operator - a company that offers cellular service, but which does not own its own cellular network. MVNOs lease capacity from the major carriers, and are often actually owned by the larger carrier as well.

Network A collection of devices connected together is called a network.

See also: "LAN," "WAN," and "Cell"

Operator

Operator is another label often used to describe cellular companies, since they are operating a network on behalf of their customers.

The network operator owns or controls the licensed radio spectrum and the network infrastructure necessary to provide service over that spectrum.

Contrast this to an MVNO, which does not own the network but which leases service from a network operator.

See also: "Carrier," "MVNO"

Oscillation

If the outside antenna for a cellular booster picks up the signal from the inside antenna, oscillation happens; if the booster is properly designed, it will shut itself down.

This is the same phenomena as walking too close to a speaker with a microphone – leading to a howling screech.

The best way to avoid oscillation is to put as much distance as possible between the inside and outside booster antennas, and to keep the antennas pointed away from each other.

Ping

How much time elapses before receiving a response to a signal is the ping time.

See also: "Latency"

POE (aka Power Over Ethernet)

Some networking equipment (such as the roof-mounted WiFiRanger units) are powered over the ethernet network wire instead of via a dedicated power cord and power supply. This makes wiring much simpler, since only a single wire needs to be run to the device. To get the power onto the ethernet wire, a POE injector is used to energize the wire. The WiFiRanger Go has a built in POE injector on one of its ethernet ports.

POTS

POTS is shorthand for Plain Old Telephone Service – the old way of getting online via dialing a modem on a regular wired phone line. Also known as a landline.

Refarming

When cellular carriers shut down older networks, they free up frequency spectrum that they can then reuse to support newer technologies. This process is called refarming the network.

The downside of refarming is that devices based upon older technologies will get slower and less coverage, and eventually become useless. But newer technologies are much faster and more efficient, making things (eventually) better for everyone.

217

Repeater (WiFi & Cellular)	A booster amplifies a signal - including any background noise. In contrast, a repeater actually recreates and broadcasts a new signal without the noise. Cellular repeaters are rare since to rebroadcast (and not just amplify) a signal requires close cooperation with the carrier. But WiFi range extending systems typically are repeaters.
Roaming	Roaming is when a cellular carrier has agreements with other networks to utilize their towers, helping the carrier provide connectivity to customers who are just passing through areas the carrier doesn't directly service. Though customers are rarely charged directly for roaming anymore, behind the scenes roaming costs the guest carrier a substantial amount.
Router	A router acts as the hub of a local network. The WAN (wide area network) connection from the router provides the upstream connection from your local network to the internet. Usually a router is required to allow multiple devices to share a single internet connection.
RSSI	RSSI stands for Received Signal Strength Indication – an arbitrary mapping of the power received by an antenna (in dBm) to a number. How this number is calculated varies greatly from device to device. This is usually simplified into a visual bars or dots display.
Satellite	Satellites in orbit above the earth are invaluable for long-distance communications. Some satellites are geosynchronous, which means they sit directly over a fixed point on the equator and are always in the exact same position in the sky. But because they are so far away, precise aiming with a dish-style antenna is essential. Others satellites are deployed in "constellations" with multiple satellites in a lower orbit working together, with the goal being that at least one satellite in the constellation is always in view. Because low earth orbit satellites are much closer and are in constant motion, aiming at them is not practical nor needed.

218

SIM

SIM stands for Subscriber Identity Module. The tiny SIM card is mandated on GSM and LTE networks, and identifies you to the network.

Swapping your SIM into a new phone essentially moves your service and phone number to that new device.

SIMs come in a range of sizes – Full-Size is extinct, but Mini-SIM, Micro-SIM, and Nano-SIM are all in common circulation. Because the Mini-SIM was the standard for so long, that size is often referred to as "full sized" or "standard sized."

Using cutters and adaptors, you can transplant larger SIM cards into devices that have smaller SIM slots, and vice-versa.

SIM cards are often tricky to find but can usually be found behind the removable battery of some devices, or in small tray that can be ejected from the side of a phone with a pin.

Skype

A cross-platform audio and video chat platform now owned by Microsoft. Skype is probably the most widely used video calling system.

See also: "FaceTime" and "Hangouts"

Smartphone

Phones that are actually computers running a general purpose operating system that can be extended with applications are called smartphones. Phones lacking an extensible operating system are called feature phones. Apple's iOS (used in the iPhone) and Google's Android are the two dominant smartphone operating systems.

SMS (aka texting, txt)

SMS stands for Short Message Service – it is a standard technology for sending text messages of 160 characters or shorter directly between cellular devices.

Spectrum

The electromagnetic spectrum is the range of all possible frequencies of radiation, ranging from extremely high-frequency gamma rays and X-rays through visible light to infrared to radio waves.

The radio spectrum is defined as 300GHz to 3kHz, and it is broken for convenience into bands.

See also: "Frequency," "Band"

Sprint Spark Sprint's marketing term for its tri-band LTE network. Spark devices use 800MHz frequencies for long range, 1900MHz for medium range, and 2500MHz for enhanced speeds and capacities in major urban areas.

Tethering Generally, tethering refers to using a USB cable to share a cellular device's connection with a laptop or router. This is the wired version of creating a personal hotspot.

Throttling Throttling is the act of intentionally slowing down a cellular data connection to run at a slower speed. Some carriers have moved away from charging overage charges when you hit your monthly data limits and instead now throttle users down to slower speeds.

In some cases, the throttling can be severe – turning a broadband LTE connection capable of 50Mbps into a snail-speed connection struggling to deliver 128Kbps.

"Unlimited" Data Read the fine print and beware of throttling. Because even "truly unlimited" rarely actually is.

VOIP VOIP stands for Voice Over Internet Protocol, and refers to technology and service providers that allow for traditional phone calls to be placed over the internet. VOIP calls can be placed either via specialized applications or by plugging a regular landline phone into a VOIP adaptor that connects to the internet.

VoLTE (aka Voice over LTE) Right now, to make a voice phone call, cell phones switch their radios back to the old voice network to make the call. This is why some Verizon phones can't surf the web or get email while a call is underway, and AT&T phones drop back to 3G data speeds until the call is completed.

Voice over LTE changes things up by sending your voice call as data over the 4G/LTE data network. This keeps your data connection running full speed and allows for HD Voice, which will let your voice calls to other VoLTE phones come through sounding more CD-quality than AM radio.

VoLTE also should make it possible to seamlessly switch between voice and video calls, since it is all just data flowing now – there is no separate "call" mode unique to old-school voice calls.

VoLTE (aka Voice over LTE) **(continued)**	VoLTE has a lot of advantages – and it opens the doors to the carriers eventually being able to fully retire their legacy voice networks someday down the road – freeing up a lot more space for fast data.
	As of early 2015, T-Mobile and Verizon support VoLTE on compatible phones nationwide, AT&T has deployed VoLTE to some markets, and Sprint has not yet announced VoLTE plans.
VPN (aka Virtual Private Network)	A VPN encrypts all the data coming to and from your computer, and sends it to a remote VPN server that then connects to the public internet on your behalf. This prevents anyone on your local network from being able to eavesdrop on you in any way – they can't even tell what sites you visit.
	If you are traveling internationally, you can use a VPN server so that you appear still to be connected from your home country. Sites that block international traffic (like Netflix streaming) will still work, and other sites will continue to default to your native language.
	A VPN will, however, introduce a small delay in all your network connections.
WAN	WAN stands for Wide Area Network, as opposed to LAN which stands for Local Area Network. Routers often have an ethernet port labeled WAN, and this is where you connect the upstream network if you have a cable or DSL modem.
Wideband LTE	T-Mobile is labeling the places where they have deployed extra large 15MHz or more bandwidth LTE channels as Wideband LTE, which should be comparable to what Verizon is calling XLTE.
	As of December 2014, T-Mobile has 121 Wideband LTE cities.
WiFi	WiFi is the popular name for local area wireless networking technology. The technical standards that define WiFi and which ensure interoperability are the 802.11 standard. Typical WiFi networks can communicate several hundred feet.
WiFi as WAN	Some routers support using a WiFi network as their upstream data connection, and this is called WiFi as WAN. With WiFi as WAN, you can use your router to connect to campground WiFi while still maintaining a private local wired and wireless network.

WiFi Calling (aka VoWiFi)

Some phones and carriers support making and receiving voice calls over a WiFi connection as well as over a cellular connection.

T-Mobile in particular has made this a core feature on their network, helping to compensate for their relatively limited cellular coverage.

WiMAX

WiMAX is a competing 4G networking standard that for a while was seen as a potential rival to LTE, and was Sprint's initial 4G network of choice.

LTE won out in the end, and all the major carriers have converged on LTE as their networking technology of the future.

Sprint has stopped expanding its WiMAX network and will be shutting it down entirely by the end of 2015 to free up more spectrum for LTE.

The final shutdown of Sprint's WiMAX network is slated for November 6th, 2015.

Some fixed-location wireless internet service providers (WISPs) still use WiMAX to provide service – it is very well suited to this sort of use.

WISP (aka Wireless ISP)

The term WISP (wireless internet service provider) has typically been attached to ISPs that deliver service to fixed locations via long-range wireless signals, often in rural areas where cable and DSL are not practical options.

A WISP providing service may have a transmitter on a nearby tower or mountain, and then install directional antennas towards it on the sides of customers' buildings. RVers in fixed locations can sometimes take advantage of a WISP for fast unlimited connectivity.

XLTE

Verizon's extended LTE network, now live in many major metro areas.

The Ongoing Conversation

Introducing: Mobile Internet Aficionados

If this book is the textbook..

...the MIA is the classroom.

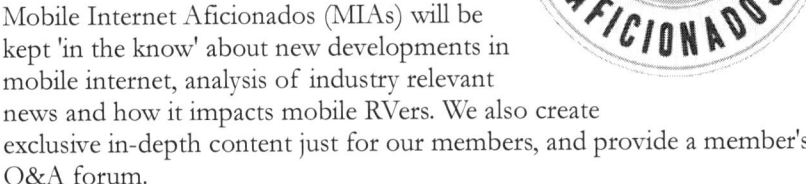

Mobile Internet Aficionados (MIAs) will be kept 'in the know' about new developments in mobile internet, analysis of industry relevant news and how it impacts mobile RVers. We also create exclusive in-depth content just for our members, and provide a member's Q&A forum.

We'll keep on top of the industry, alert you to anything that might impact you and answer your questions - so you can concentrate on what drives you!

This technology stuff changes often. It's almost guaranteed that as soon as we submit the manuscript to this book to be published, things will change. A new product will be released, a plan will be changed, or an announcement will be made.

With the 2014 release of this book via our funding campaign, we also launched this premium membership service to keep our members "in the know". The service has been met with great enthusiasm and success, and is our pure joy to offer.

This service is designed for those who depend on mobile internet to enable their roaming lifestyle but don't have the time to keep on top of it.

We'll scour news sources and forums and keep in touch with our industry contacts – and write an analysis on what it means to us mobile folks.

Via this service, you'll have access to:

- Our exclusive member's forums – where you can ask us questions and get answers.
- Our exclusive membership newsletter – where we'll analyze mobile technology news and report to you what you need to know, as well as alert you to any upcoming changes that might affect you.

223

- Exclusive in-depth content released to members first - such as reviews, equipment testing results and guides. All content on the site is always advertising free too for members.

- Access to exclusive interactive webinars.

- Discounts on private mobile internet advising sessions with us.

- *Free Updates to this book* in PDF format.

If this stuff is vital to your mobile livelihood, we invite you to consider joining us at:

www.RVMobileInternet.com/membership

And even if you don't join Mobile Internet Aficionados, do check the site from time to time – we will continue to post general public and free updates there that might impact decisions you make about mobile internet.

Save $5 on Membership

Our Mobile Internet Aficionados premium membership includes a copy of this book in PDF eBook format, and any updates that might be issued during your membership term.

To thank you for already having purchased this book separately, we offer you $5 off the price of a new annual membership to the MIA.

When joining at www.RVMobileInternet.com/membership, select the full Mobile Internet Aficionado package membership. At checkout use the below code to save $5:

Join the MIA

Save $5 off a new Annual Membership

Use Coupon Code:

IBoughtTMIH2015P

224

Private Mobile Internet Advising Sessions

Ever wished you could just ask someone what you need to keep connected on the road? Well, now you can.

Everyone's needs are unique, and highly dependent upon how important internet is to you, your budget and your travel style. We'd love to help you navigate this stuff, and offer personal advising sessions so we can take the time to get to know you and your needs to make an informed recommendation.

We'll be happy to book a session with you to discuss your goals of keeping online while being mobile and help you assemble an ideal mobile internet arsenal custom designed just for you. We offer private sessions for those who would like some additional help in figuring out their ideal solution.

We start with having you fill out an assessment interview, schedule a phone or video session, and then follow up with a custom written report with our specific recommendations.

For more information or to schedule a session:

www.RVMobileInternet.com/advising

RV Mobile Internet Resource Center

Keep in the know as this stuff changes... for news, further articles, resources and updated editions of this book visit:

www.RVMobileInternet.com

Join our **FREE** e-mail newsletter for monthly updates on what's changed in the mobile internet landscape (we'll also alert you to new editions of this book coming out.):

http://eepurl.com/0KJG1

You are also welcome to join our free public Facebook group for discussions with other RVers (ourselves included) interested in this topic:

www.facebook.com/groups/rvinternet